Student Solutions Manual

for use with

Practical Business Math Procedures

Seventh Edition

Jeffrey Slater
North Shore Community College

Boston Burr Ridge, IL Dubuque, IA Madison, WI New York San Francisco St. Louis
Bangkok Bogotá Caracas Kuala Lumpur Lisbon London Madrid Mexico City
Milan Montreal New Delhi Santiago Seoul Singapore Sydney Taipei Toronto

McGraw-Hill Higher Education

A Division of The McGraw·Hill Companies

Student Solutions Manual for use with
PRACTICAL BUSINESS MATH PROCEDURES
Jeffrey Slater

3 4 5 6 7 8 9 0 BKM/BKM 0 9 8 7 6 5 4 3

ISBN 0-07-253737-X

www.mhhe.com

Preface

This solutions booklet is intended to provide a "back up" for your study of business math. The manual contains step-by-step solutions for an example or two of every kind of problem in the text. It also contains a student progress chart that lets you keep a record of your progress.

Please realize that the worked out solutions should be used to help, not replace your effort at solving the problems. Here are a few tips to keep in mind while doing your assignments.

1. Put this Solutions Manual on another table where you can see it, but where you can't reach it. Then think of trying to "beat it" by solving the problem without getting up to refer to it.

2. When you get to a tough problem and just can't quite solve it, go to another problem. Does it make sense? If not, go back and reread the problem.

3. If you are still stuck, try a guess and substitute the answer back into the problem. Does it seem logical? If not, go back and reread the problem remembering whether your guess was close or not.

4. Try looking at the example in the text that is similar to the problem. Does the method in the book give you any clues? Now look at the steps you wrote down in trying to solve the problem. How does it compare to the example?

5. When you do look up the solution in this manual, cover the answer from the so that only the first line shows. Then look at the first step only. How does it compare to your first step? Then go down line by line.

6. Just stick with it. You'll do fine!

Contents

Student Progress Chart vii

Chapter 1: Whole Numbers: How to Dissect and Solve Word Problems 1

Chapter 2: Fractions 5

Chapter 3: Decimals 9

Chapter 4: Banking 13

Chapter 5: Solving for the Unknown: A How-To Approach for Solving Equations 17

Chapter 6: Percents and Their Applications 21

Chapter 7: Discounts: Trade and Cash 25

Chapter 8: Markups and Markdowns: Insight into Perishables 29

Chapter 9: Payroll 33

Chapter 10: Simple Interest 37

Chapter 11: Promissory Notes, Simple Discount Notes, and the Discount Process 41

Chapter 12: Compound Interest and Present Value 43

Chapter 13: Annuities and Sinking Funds 45

Chapter 14: Installment Buying, Rule of 78, and Revolving Charge Credit Cards 49

Chapter 15: The Cost of Home Ownership 53

Chapter 16: How to Read, Analyze, and Interpret Financial Reports 55

Chapter 17: Depreciation 59

Chapter 18: Inventory and Overhead 61

Chapter 19: Sales, Excise, and Property Taxes 63

Chapter 20: Life, Fire, and Auto Insurance 65

Chapter 21: Stocks, Bonds, and Mutual Funds 67

Chapter 22: Business Statistics 69

Contents

Chapter 9 ....

Chapter 10 ....

Chapter 11 ....

Chapter 12 ....

Chapter 13 ....

Chapter 14 ....

Chapter 15 ....

Chapter 16 ....

Chapter 17 ....

Chapter 18 ....

Chapter 19 ....

Chapter 20 Safe, Secure, and Clean ....

Chapter 21 ....

Chapter 22 The Last Chapter ....

STUDENT PROGRESS CHART

Chapter	LU	Title	Objectives reviewed	Practice quizzes completed	Review chapter** organizer with vocabulary and drill and word problems completed	Challenge problems completed	Summary practice test completed	Video and/or computer software reviewed	Additional set of drill and word problems by learning unit or cumulative review completed	Exam grade on chapter
1	1-1	Reading/Writing Whole Numbers								
	1-2	Addition/Subtraction								
	1-3	Multiplication/Division								
2	2-1	Fractions/Conversions								
	2-2	Fractions/Add/Sub								
	2-3	Fractions/Mult/Div								
3	3-1	Dec/Conversions/Rounding								
	3-2	Dec/Add/Sub/Mult/Div/Foreign Currency								
4	4-1	Checking Account/Credit Cards								
	4-2	Bank Reconciliation								
5	5-1	Equations								
	5-2	Word Problems								
6	6-1	Percent/Conversions								
	6-2	Portion Formula								
7	7-1	Trade Discounts								
	7-2	Cash Discount/Due Dates/Freight								
8	8-1	Markup-Cost								
	8-2	Markup-Selling Price								
	*8-3	Markdowns/Perishables								
9	9-1	Payroll-Gross								
	9-2	Payroll Deductions/Employer Responsibilities								
10	10-1	Simple Interest								
	10-2	Principal/Rate/Time								
	10-3	U.S. Rule								
11	11-1	Promissory/Simple Discount Note								
	11-2	Discounting								
12	12-1	Compound Interest								
	12-2	Present Value								

* Not in Brief Seventh Edition.; **Self-paced learning worksheets in Business Math Handbook/Study Guide could be complete.

STUDENT PROGRESS CHART

Chapter	LU	Title	Objectives reviewed	Practice quizzes completed	Review chapter** organizer with vocabulary and drill and word problems completed	Challenge problems completed	Summary practice test completed	Video and/or computer software reviewed	Additional set of drill and word problems by learning unit or cumulative review completed	Exam grade on chapter
13	13-1	Annuities/Ordinary/Due								
	13-2	Present Value of Annuity								
	13-3	Sinking Fund								
14	14-1	Installment Buying								
	14-2	Rule of 78								
	14-3	Revolving Charge/Av. Daily Balance								
15	15-1	Monthly Payment								
	15-2	Amortization Schedule								
16	16-1	Balance Sheet								
	16-2	Income Statement								
	16-3	Trend/Ratio Analysis								
17	17-1	Concepts/Straight Line								
	17-2	Units-of-Production								
	17-3	Sum-of-the Years'-Digits								
	17-4	Declining-Balance								
	17-5	MACRS								
18	18-1	Sp. Id./Wtd. Av./FIFO/LIFO								
	18-2	Retail/Gross Profit/Turn/Overload								
19	19-1	Sales/Excise Tax								
	19-2	Property Tax								
20	20-1	Life Insurance								
	20-2	Fire Insurance								
	20-3	Auto Insurance								
21	21-1	Stocks								
	21-2	Bonds								
22	22-1	Mean/Median/Mode								
	22-2	Frequency Distribution/Graphs								
	22-3	Measures of Dispersion								

END-OF-CHAPTER PROBLEMS

DRILL PROBLEMS

Subtract the following:

1–9.
$$
\begin{array}{r}
80 \\
-42 \\
\end{array}
\qquad
\begin{array}{r}
^{7}\!|^{10} \\
8\cancel{0} \\
-42 \\
\hline
38
\end{array}
$$

1–13.
$$
\begin{array}{r}
1,622 \\
-548 \\
\hline
1,074
\end{array}
$$

Multiply the following:

1–15.
$$
\begin{array}{r}
510 \\
\times\;61 \\
\hline
510 \\
30\;60 \\
\hline
31,110
\end{array}
$$

1–19.
$$
\begin{array}{r}
450 \\
\times\;280 \\
\hline
36\;000 \\
90\;0 \\
\hline
126,000
\end{array}
$$

Divide the following by short division:

1–21. $9\overline{)810}$ with quotient 90

Divide the following by long division. Show work and remainder.

1–23.
$$
\begin{array}{r}
86\,\text{R4} \\
6\overline{)520} \\
48 \\
\hline
40 \\
36 \\
\hline
4
\end{array}
$$

1–29. Add the following and check by totaling each column individually without carrying numbers:

	Check
8,539	16
6,842	16
+ 9,495	17
24,876	23
	24,876

Estimate the following by rounding all the way and then do actual addition:

	Actual	**Estimate**
1–31.	6,980	7,000
	3,190	3,000
	+ 7,819	+ 8,000
	17,989	18,000

Divide the following and check by multiplication:

1–39.
$$
\begin{array}{r}
19\,\text{R21} \\
45\overline{)876} \\
45 \\
\hline
426 \\
405 \\
\hline
21
\end{array}
$$

Check

$45 \times 19 = 855$

$$
\begin{array}{r}
855 \\
+ 21(\text{R}) \\
\hline
876
\end{array}
$$

1-41. Add the following columns horizontally and vertically:

Production Report						
	Monday	**Tuesday**	**Wednesday**	**Thursday**	**Friday**	
Software packages	450	92	157	24	40	= 763
Laptops	490	75	44	77	30	= 716
Video	325	82	22	44	18	= 491
Computer monitors	66	24	51	66	50	= 257
	1,331	+273	+274	+211	+138	= 2,227

Using data in Problem 1–41, answer the following:

1-43. If two weeks ago production was 7 times the total of this report, what was total production?

$2{,}227 \times 7 = 15{,}589$

1-49. Estimate actual problem by rounding all the way and do actual division:

Actual

$$
\begin{array}{r}
12\ \text{R}610 \\
695\overline{)8{,}950} \\
6\ 95 \\
\hline
2\ 000 \\
1\ 390 \\
\hline
610
\end{array}
$$

Estimate

$$
\begin{array}{r}
12\ \text{R}600 \\
700\overline{)9{,}000} \\
7\ 00 \\
\hline
2\ 000 \\
1\ 400 \\
\hline
600
\end{array}
$$

WORD PROBLEMS

1-51. On May 22, 2001, *USA Today* compared the amount of money Americans spent on playing state lotteries. A total of thirty-eight billion dollars was spent in 2000. Massachusetts topped the list with lottery sales of three billion, seven hundred thousand dollars. Montana was at the bottom, spending twenty-nine million dollars. In numerical form, how much more did Massachusetts residents spend compared to Montana residents?

$$
\begin{array}{r}
\overset{2\ 9\ 9\ 10}{\$3{,}\cancel{0}\cancel{0}\cancel{0}{,}700{,}000}} \\
-\quad\ \ 29{,}000{,}000 \\
\hline
\$2{,}971{,}700{,}000 \text{ more}
\end{array}
$$

1-55. NTB Tires bought 910 tires from its manufacturer for $36 per tire. What is the total cost of NTB's purchase? If the store can sell all the tires at $65 each, what will be the store's gross profit, or the difference between its sales and costs (Sales − Costs = Gross profit)?

Cost = 910 × $36 = $32,760   Sales = 910 × $65 = $59,150

$$
\begin{array}{r}
\$59{,}150 \text{ sales} \\
-\ 32{,}760 \text{ cost} \\
\hline
\$26{,}390 \text{ gross profit}
\end{array}
$$

1-57. Jose Gomez bought 4,500 shares of Microsoft stock. He held the stock for 6 months. Then Jose sold 180 shares on Monday, 270 shares on Tuesday and again on Thursday, and 800 shares on Friday. How many shares does Jose still own? The average share of the stock Jose owns is worth $52 per share. What is the total value of Jose's stock?

$$
\begin{array}{r}
4{,}500 \text{ shares bought} \\
-1{,}520 \text{ shares sold} \\
\hline
2{,}980
\end{array}
$$

$180 + 270 + 270 + 800 = 1{,}520$

2,980 shares × $52 = $154,960

1-61. Ron Alf, owner of Alf's Moving Company, bought a new truck. On Ron's first trip, he drove 1,200 miles and used 80 gallons of gas. How many miles per gallon did Ron get from his new truck? On Ron's second trip, he drove 840 miles and used 60 gallons. What is the difference in miles per gallon between Ron's first trip and his second trip?

$1{,}200 \div 80 = 15$ miles per gallon

$840 \div 60 = 14$ miles per gallon   Difference = 1 mile per gallon

2

1–66. Hometown Buffet had 90 customers on Sunday, 70 on Monday, 65 on Tuesday, and a total of 310 on Wednesday to Saturday. How many customers did Hometown Buffet serve during the week? If each customer spends $9, what were the total sales for the week?

$$90 + 70 + 65 + 310 = \quad 535 \text{ customers}$$
$$\times \quad \$9$$
$$\overline{\$4,815}$$

If Hometown's Buffet had the same sales each week, what were the sales for the year?
$4,815 \times 52 = \$250,380$

1–73. MVP, an athletic sports shop, bought and sold the following merchandise:

	Cost	Selling price
Tennis rackets	$ 2,900	$ 3,999
Tennis balls	70	210
Bowling balls	1,050	2,950
Sneakers	+ 8,105	+ 14,888
	$12,125	$22,047

What was the total cost of merchandise bought by MVP? If the shop sold all its merchandise, what were the sales and the resulting gross profit (Sales − Costs = Gross profit)?

Sales	$22,047
−Costs	− 12,125
=Gross profit	$ 9,922

1–77. While redecorating, Paul Smith went to Home Depot and bought 125 square yards of commercial carpet. The total cost of the carpet was $3,000. How much did Paul pay per square yard?
$3,000 \div 125 = \$24$ per square yard

CHALLENGE PROBLEM

1–79. On June 1, 2001, *USA Today* compared financial contributions that the tobacco industry gave to political parties in the 1999–2000 election. The total dollar amount was eight million, four hundred thousand, with the following companies in the top five:

1. Philip Morris—three million, four hundred fifty thousand, one hundred thirty-nine.
2. UST Inc.—one million, five hundred eighty-eight thousand, three hundred fifty-four.
3. RJ Reynolds Tobacco—nine hundred ninety-one thousand, four hundred twenty-seven.
4. Brown & Williamson Tobacco—nine hundred seventy-nine thousand, seven hundred thirty-two.
5. Loews Corp.—two hundred eighty-five thousand, fifty.

(a) In verbal form, what was the total dollar amount contributed by the top five tobacco companies?
(b) In numerical form, what is the difference between the total dollar amount contributed by the tobacco industry and the dollar amount contributed by the top five tobacco companies? **(c)** What was Philip Morris's average monthly contribution? Round your answer to the nearest hundred thousands.

a. $3,450,139
    1,588,354
     991,427
     979,732
+   285,050
$7,294,702 numerical total

= Seven million, two hundred ninety-four thousand, seven hundred two

b. $8,400,000 industry total
− 7,294,702 top five
$1,105,298 difference

c. $287,511 = $300,000
12)$3,450,139
2 4
1 05
96
90
84
61
60
13
−12
19
−12
R7

END-OF-CHAPTER PROBLEMS

DRILL PROBLEMS

Identify the following types of fractions:

2–3. $\dfrac{16}{11}$ Improper

Convert the following to mixed numbers:

2–5. $\dfrac{921}{15} = 61\dfrac{6}{15} = 61\dfrac{2}{5}$

Reduce the following to the lowest terms. Show how to calculate the greatest common divisor by the step approach.

2–9. $\dfrac{44}{52} = \dfrac{44 \div 4}{52 \div 4} = \dfrac{11}{13}$

$$44\overline{)52} \quad 8\overline{)44} \quad 4\overline{)8}$$
$$\underset{8}{\underline{44}} \quad\nearrow\quad \underset{4}{\underline{40}} \quad\nearrow\quad \underset{0}{\underline{8}}$$

Determine the LCD of the following **(a)** by inspection and **(b)** by division of prime numbers:

2–13. $\dfrac{1}{4}, \dfrac{3}{32}, \dfrac{5}{48}, \dfrac{1}{8}$

Inspection   96
$2 \times 2 \times 2 \times 2 \times 2 \times 3 = 96$

Check

2 /	4	32	48	8
2 /	2	16	24	4
2 /	1	8	12	2
2 /	1	4	6	1
	1	2	3	1

Subtract the following and reduce to lowest terms:

2–21. $12\dfrac{1}{9} - 4\dfrac{2}{3}$

$$12\dfrac{1}{9} = 11\dfrac{10}{9} \quad \left(\dfrac{9}{9} + \dfrac{1}{9}\right)$$
$$-4\dfrac{6}{9} = -4\dfrac{6}{9}$$
$$\overline{7\dfrac{4}{9}}$$

Multiply the following. Use the cancellation technique.

2–25. $\dfrac{4}{10} \times \dfrac{30}{60} \times \dfrac{6}{10} = \dfrac{\overset{1}{\cancel{\overset{2}{\cancel{4}}}}}{\underset{5}{\cancel{10}}} \times \dfrac{\overset{3}{\cancel{30}}}{\underset{\underset{5}{10}}{\cancel{60}}} \times \dfrac{\overset{1}{\cancel{6}}}{\underset{1}{\cancel{10}}} = \dfrac{3}{25}$

WORD PROBLEMS

2–33. U.S. Airways pays Paul Lose $125 per day to work in security at the airport. Paul became ill on Monday and went home after $\frac{1}{5}$ of a day. What did he earn on Monday? Assume no work, no pay.

$$\dfrac{1}{5} \times \$125 = \dfrac{\$125}{5} = \$25$$

2–37. The June 2001 *Woodsmith* magazine tells how to build a country wall shelf. Two side panels are $\frac{3}{4} \times 7\frac{1}{2} \times 31\frac{5}{8}$ inches long. **(a)** What is the total length of board you will need? **(b)** If you have a board $74\frac{1}{3}$ inches long, how much of the board will remain after cutting?

a. $31\frac{5}{8}$ inches $\times 2 = \dfrac{253}{\overset{}{\underset{4}{8}}} \times \dfrac{\overset{1}{2}}{1} = \dfrac{253}{4} = 63\frac{1}{4}$ inches

b.
$$74\frac{1}{3} = \quad 74\frac{4}{12}$$
$$-\,63\frac{1}{4} = \; -\,63\frac{3}{12}$$
$$\overline{\qquad\qquad 11\frac{1}{12} \text{ inches remain}}$$

2–43. Ajax Company charges \$150 per cord of wood. If Bill Ryan orders $3\frac{1}{2}$ cords, what will his total cost be?

$$\$150 \times 3\frac{1}{2} = \overset{\$75}{\cancel{\$150}} \times \dfrac{7}{\underset{1}{2}} = \$525$$

2–45. Marc, Steven, and Daniel entered into an Internet partnership. Marc owns $\frac{1}{9}$ of the Dot.com, and Steven owns $\frac{1}{4}$. What part does Daniel own?

$$\dfrac{4}{36} + \dfrac{9}{36} = \dfrac{13}{36} \qquad\qquad 1 - \dfrac{13}{36} = \dfrac{23}{36} \text{ for Daniel or } \dfrac{36}{36} - \dfrac{13}{36} = \dfrac{23}{36}$$

2–49. Michael, who loves to cook, makes apple cobbler (serves 6) for his family. The recipe calls for $1\frac{1}{2}$ pounds of apples, $3\frac{1}{4}$ cups of flour, $\frac{1}{4}$ cup of margarine, $2\frac{3}{8}$ cups of sugar, and 2 teaspoons of cinnamon. Since guests are coming, Michael wants to make a cobbler that will serve 15 (or increase the recipe $2\frac{1}{2}$ times). How much of each ingredient should Michael use?

$$\dfrac{3}{2} \times \dfrac{5}{2} = \dfrac{15}{4} = 3\frac{3}{4} \text{ pounds of apples} \qquad\qquad \dfrac{19}{8} \times \dfrac{5}{2} = \dfrac{95}{16} = 5\frac{15}{16} \text{ cups of sugar}$$

$$\dfrac{13}{4} \times \dfrac{5}{2} = \dfrac{65}{8} = 8\frac{1}{8} \text{ cups of flour} \qquad\qquad 2 \times \dfrac{5}{2} = 5 \text{ teaspoons of cinnamon}$$

$$\dfrac{1}{4} \times \dfrac{5}{2} = \dfrac{5}{8} \text{ cup of margarine}$$

2–55. McGraw-Hill/Irwin publishers stores some of its inventory in a warehouse that has 14,500 square feet of space. Each book requires $2\frac{1}{2}$ square feet of space. How many books can McGraw-Hill/Irwin keep in this warehouse?

$$14,500 \div 2\frac{1}{2} = \overset{2,900}{\cancel{14,500}} \times \dfrac{2}{\underset{1}{\cancel{5}}} = 5,800 \text{ books}$$

ADDITIONAL SET OF WORD PROBLEMS

2–59. On July 2, 2001, you received a special ad from Home Depot stating that a $\frac{3}{4}$ inch $\times$ 10 feet piece of PVC piping is on sale for \$1.39. You plan to install the PVC piping in your basement. The measurements you have calculated include pieces with a total length of $11\frac{3}{4}$ feet, $15\frac{3}{8}$ feet, and $8\frac{5}{16}$ feet. **(a)** What is the total length of piping needed? **(b)** If you purchased a total of 40 feet of piping, how much piping will you have left over?

a.
$$11\frac{3}{4} \text{ feet} = 11\frac{12}{16}$$
$$+\,15\frac{3}{8} \quad= 15\frac{6}{16}$$
$$+\,8\frac{5}{16} \quad= \;8\frac{5}{16}$$
$$\overline{\qquad 34\frac{23}{16} = 35\frac{7}{16} \text{ feet}}$$

b.
$$40 \quad= \quad 39\frac{16}{16}$$
$$-\,35\frac{7}{16} = \;-\,35\frac{7}{16}$$
$$\overline{\qquad\qquad 4\frac{9}{16} \text{ feet left}}$$

6

2–61. Frank Puleo bought 6,625 acres of land in ski country. He plans to subdivide the land into parcels of $13\frac{1}{4}$ acres each. Each parcel will sell for $125,000. How many parcels of land will Frank develop? If Frank sells all the parcels, what will be his total sales?

$$6{,}625 \div 13\frac{1}{4} = 6{,}625 \times \frac{4}{53} = 500 \text{ parcels} \times \$125{,}000 = \$62{,}500{,}000$$

If Frank sells $\frac{3}{5}$ of the parcels in the first year, what will be his total sales for the year?

$$\frac{3}{\overset{}{\underset{1}{5}}} \times \overset{100}{\cancel{500}} = 300 \times \$125{,}000 = \$37{,}500{,}000$$

CHALLENGE PROBLEM

2–69. Jack MacLean has entered into a real estate development partnership with Bill Lyons and June Reese. Bill owns $\frac{1}{4}$ of the partnership, while June has a $\frac{1}{5}$ interest. The partners will divide all profits on the basis of their fractional ownership.

The partnership bought 900 acres of land and plans to subdivide each lot into $2\frac{1}{4}$ acres. Homes in the area have been selling for $240,000. By time of completion, Jack estimates the price of each home will increase by $\frac{1}{3}$ of the current value. The partners sent a survey to 12,000 potential customers to see whether they should heat the homes with oil or gas. One-fourth of the customers responded by indicating a 5-to-1 preference for oil. From the results of the survey, Jack now plans to install a 270-gallon oil tank at each home. He estimates that each home will need 5 fills per year. Current price of home heating fuel is $1 per gallon. The partnership estimates its profit per home will be $\frac{1}{8}$ the selling price of each home.

From the above, please calculate the following:

a. Number of homes to be built.

$$900 \div 2\frac{1}{4} = \overset{100}{\cancel{900}} \times \frac{4}{\underset{1}{\cancel{9}}} = 400 \text{ homes}$$

b. Selling price of each home.

$$1\frac{1}{3} \times \$240{,}000 = \frac{4}{\underset{1}{\cancel{3}}} \times \overset{\$80{,}000}{\cancel{\$240{,}000}} = \$320{,}000$$

c. **(1).** Number of people responding to survey.

$$\frac{1}{4} \times 12{,}000 = 3{,}000 \text{ people}$$

(2). Number of people desiring oil.

$$\frac{5}{6} \times 3{,}000 = 2{,}500 \text{ people}$$

d. Average monthly cost to run oil heat per house.

$$270 \times 5 = 1{,}350 \times \$1 = \frac{\$1{,}350}{12} = \$112.50$$

e. Amount of profit Jack will receive from the sale of homes.

$$\frac{1}{4} + \frac{1}{5} = \frac{5}{20} + \frac{4}{20} = \frac{9}{20} \qquad 1 - \frac{9}{20} = \frac{11}{20} \text{ for Jack}$$

$$\frac{1}{\underset{1}{\cancel{8}}} \times \overset{\$40{,}000}{\cancel{\$320{,}000}} = \$40{,}000$$

$$\begin{array}{r} \$40{,}000 \\ \times\ 400 \\ \hline \$16{,}000{,}000 \end{array}$$

$$\frac{11}{20} \times \$16{,}000{,}000 = \$8{,}800{,}000$$

END-OF-CHAPTER PROBLEMS

DRILL PROBLEMS

Round the following as indicated;

		Tenth	Hundredth	Thousandth
3–3.	.9482	.9	.95	.948

Convert the following types of decimal fractions to decimals (round to nearest hundredth as needed):

3–17. $14\frac{91}{100}$ 14.91

Convert the following decimals to fractions. Do not reduce to lowest terms.

3–25. .7065 $\frac{7,065}{10,000}$

Rearrange the following and add:

3–33. .115, 10.8318, 4.7, 802.4811
       818.1279

Convert the following to decimals and round to the nearest hundredth:

3–51. $\frac{5}{8}$ .63

WORD PROBLEMS

As needed, round answers to the nearest cent.

3–67. The stock of Intel has a high of $30.25 today. It closed at $28.85. How much did the stock drop from its high?

$$\begin{array}{r} {\scriptstyle 9\ \ 12} \\ \$30.25 \\ -\ 28.85 \\ \hline \$\ 1.40 \end{array}$$

3–71. Pete Allan bought a scooter on the Web for $99.99. He saw the same scooter in the mall for $108.96. How much did Pete save by buying on the Web?

$$\begin{array}{r} {\scriptstyle 17\ 18} \\ {\scriptstyle 9\ 7\ 8\ 16} \\ \$108.96 \\ -\ 99.99 \\ \hline \$\ 8.97 \end{array}$$

3–79. Audrey Long went to Japan and bought an animation cell of Mickey Mouse. The price was 25,000 yen. What is the price in U.S. dollars? Check your answer.
       25,000 × $.00816 = $204.00
       $204.00 × 122.53 = 24,996.12 yen   (Rounded to 25,000 yen)

ADDITIONAL SET OF WORD PROBLEMS

3–81. Tie Yang bought season tickets to the Boston Pops for $698.55. The season package included 38 performances. What is the average price of the tickets per performance? Round to nearest cent. Sam, Tie's friend, offered to buy 4 of the tickets from Tie. What is the total amount Tie should receive?
       $698.55 ÷ 38 = $18.38 × 4 = $73.52

3–83. *The New York Times*, dated June 3, 2001, compared a gallon of gasoline to a gallon of diesel. Diesel currently sells at an average of $1.45 a gallon as opposed to $1.70 for gasoline. What would be the cost of a fill-up of 12.6 gallons of **(a)** diesel, **(b)** gasoline, and **(c)** what is the difference?

a.
$$\begin{array}{r} \$\ 1.45 \text{ diesel} \\ \times\ \ 12.6 \\ \hline \$18.27 \text{ total cost} \end{array}$$

b.
$$\begin{array}{r} \$\ 1.70 \text{ gasoline} \\ \times\ \ 12.6 \\ \hline \$21.42 \text{ total cost} \end{array}$$

c.
$$\begin{array}{r} \$21.42 \\ -\ 18.27 \\ \hline \$\ 3.15 \text{ difference} \end{array}$$

 CHALLENGE PROBLEM

3–87. The following items were charged in Canada to your bank credit card:

1. Domino's Pizza, London, Canada $15.32
2. Shell 3001 Dougall Ave., Windsor, Canada   25.20
3. Little Caesars, Ottawa, Canada   16.52
4. Richmond Plaza Motel, Ottawa, Canada   79.08
5. Petrocan Hwy. 401, Cambridge, Canada   14.90
6. Mr. Gas #081, Kingston, Canada   14.39
7. Days Inns, London, Canada   82.87

a. Using the text exchange rates in your *Business Math Handbook,* find the amount you should be charged for each item.

b. What should your total bill be? Check your answer.

a.
   1. $ 15.32 × .64570 = $ 9.89
   2.   25.20 × .64570 =   16.27
   3.   16.52 × .64570 =   10.67
   4.   79.08 × .64570 =   51.06
   5.   14.90 × .64570 =   9.62
   6.   14.39 × .64570 =   9.29
   7. + 82.87 × .64570 =   53.51

b. $248.28 $160.31

Check $160.31 × 1.5487 = $248.27 (off 1 cent due to rounding)

1. The top rate at the Waldorf Towers Hotel in New York is $390. The top rate at the Ritz Carlton in Boston is $345. If John spends 9 days at the hotel, how much can he save if he stays at the Ritz? *(p. 14)*

     $390
   − 345
   ‾‾‾‾‾‾‾
     $ 45 × 9 = $405

4. AT&T advertised a 10-minute call for $2.27. MCI WorldCom's rate was $2.02. Assuming Bill Splat makes forty 10-minute calls, how much could he save by using MCI WorldCom? *(p. 70)*

     $2.27
   − 2.02
   ‾‾‾‾‾‾‾
     $ .25 × 40 = $10.00

7. Lillie Wong bought 4 new Firestone tires at $82.99 each. Firestone also charged $2.80 per tire for mounting, $1.95 per tire for valves, and $3.15 per tire for balancing. Lillie turned her 4 old tires in to Firestone, which charged $1.50 per tire to dispose of them. What was Lillie's final bill? *(p. 14)*

   $$4 \times \$82.99 = \$331.96$$
   $$\$2.80 + \$1.95 + \$3.15 = \$7.90 \times 4 = \underline{31.60}$$
   $$\$363.56 + \$6.00 = \$369.56$$

9. Today the average business traveler will spend almost $50 a day on food. The breakdown is dinner, $22.26; lunch, $10.73; breakfast, $6.53; tips, $6.23; and tax, $1.98. If Clarence Donato, an executive for Honeywell, spends only $.3\overline{3}$ of the average, what is Clarence's total cost for food for the day? If Clarence wanted to spend $\frac{1}{3}$ more than the average on the next day, what would be his total cost on the second day? Round to the nearest cent. *(p. 71)*

   $22.26 + $10.73 + $6.53 + $6.23 + $1.98 = $47.73 actual

   $$\frac{1}{3} \times \$47.73 = \$15.91$$

   $$1\frac{1}{3} \times \$47.73 = \frac{4}{3} \times \$47.73 = \$63.64$$

   Be sure you use the fractional equivalent in calculating $.3\overline{3}$.

END-OF-CHAPTER PROBLEMS

DRILL PROBLEMS

4–1. Fill out the check register that follows with this information:

2004

May	8	Check No. 611	Amazon.com	$ 81.96
	15	Check No. 612	Dell Computer	33.10
	19	Deposit		800.40
	20	Check No. 613	Sprint	110.22
	24	Check No. 614	Krispy Kreme	217.55
	29	Deposit		198.10

		RECORD ALL CHARGES OR CREDITS THAT AFFECT YOUR ACCOUNT						
NUMBER	DATE 2004	DESCRIPTION OF TRANSACTION	PAYMENT/DEBIT (−)	√	FEE (IF ANY) (−)	DEPOSIT/CREDIT (+)	\$ BALANCE **1,017**	**20**
611	5/8	Amazon.com	81 96		\$	\$	− 81	96
							935	24
612	5/15	Dell Computer	33 10				− 33	10
							902	14
	5/19	Deposit				800 40	+ 800	40
							1,702	54
613	5/20	Sprint	110 22				− 110	22
							1,592	32
614	5/24	Krispy Kreme	217 55				− 217	55
							1,374	77
	5/29	Deposit				198 10	+ 198	10
							\$ 1,572	87

4–3. You are the bookkeeper of Reese Company and must complete a merchant batch header for November 10, 2004, from the following credit card transactions. The company lost the charge slips and doesn't include an adding machine tape. Reese's checking account number is 3158062. The merchant's signature can be left blank. **Credit card sales** are $210.40, $178.99, $29.30, and $82.80. **Credit card returns** are $15.10 and $22.99.

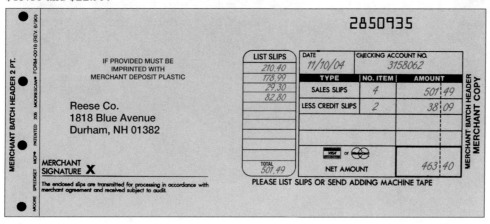

WORD PROBLEMS

4–9. The January 15, 2001, issue of the *Providence Journal* reported on Pennsylvania's banking fees. Sovereign Bancorp of Pennsylvania charges a $2.50 monthly checking fee, plus a 75-cent fee for each transaction over 10 per month. Judy Smejek has an account at Sovereign Bancorp and just received her April 3, 2002, bank statement. Included in the statement was a charge of 75 cents for 15 additional checks written. She was also charged a $2.50 service fee. The bank statement shows a $1,768.01 balance. Judy's checkbook has a $1,085.81 balance. The following checks have not cleared the bank: No. 113, $312.50; No. 114, $50.40; and No. 115, $16.80. Judy made a $650.25 deposit that is not shown on the bank statement. She has her $540 monthly mortgage payment paid through the bank. Her $1,506.50 IRS refund check was mailed to her bank. Prepare Judy Smejek's bank reconciliation.

Excel

Judy's checkbook balance		$1,085.81	Bank balance		$1,768.01
Add:			Add:		
IRS refund		1,506.50	Deposit in transit		650.25
		$2,592.31			$2,418.26
Deduct:			Deduct:		
Additional checks fee	$ 11.25		Outstanding checks:		
Mortgage automatic withdrawal	540.00		No. 113	$312.50	
Service fee	2.50	553.75	No. 114	50.40	
			No. 115	16.80	379.70
Reconciled balance		$2,038.56	Reconciled balance		$2,038.56

4–11. The *Arkansas Democrat-Gazette,* on October 29, 2000, reported that banks are finding more ways to charge fees, such as a $25 overdraft fee. Sue McVickers has an account in Fayetteville; she has received her bank statement with this $25 charge. Also, she was charged a $6.50 service fee; however, the good news is she had earned $5.15 interest. Her bank statement's balance was $315.65, but it did not show the $1,215.15 deposit she had made. Sue's checkbook balance shows $604.30. The following checks have not cleared: No. 250, $603.15; No. 253, $218.90; and No. 254, $130.80. Prepare Sue's bank reconciliation.

Sue's checkbook balance		$604.30	Bank balance		$ 315.65
Add:			Add:		
Interest		5.15	Deposit in transit		1,215.15
		$609.45			$1,530.80
Deduct:			Deduct:		
Service charge	$ 6.50		Outstanding checks:		
Overdraft	25.00	31.50	No. 250	$603.15	
			No. 253	218.90	
			No. 254	130.80	952.85
Reconciled balance		$577.95	Reconciled balance		$ 577.95

CHALLENGE PROBLEM

4–15. Melissa Jackson, bookkeeper for Kinko Company, cannot prepare a bank reconciliation. From the following facts, can you help her complete the June 30, 2004, reconciliation? The bank statement showed a $2,955.82 balance. Melissa's checkbook showed a $3,301.82 balance.

Melissa placed a $510.19 deposit in the bank's night depository on June 30. The deposit did not appear on the bank statement. The bank included two DMs and one CM with the returned checks: $690.65 DM for NSF check, $8.50 DM for service charges, and $400.00 CM (less $10 collection fee) for collecting a $400.00 non-interest-bearing note. Check No. 811 for $110.94 and check No. 912 for $82.50, both written and recorded on June 28, were not with the returned checks. The bookkeeper had correctly written check No. 884, $1,000, for a new cash register, but she recorded the check as $1,069. The May bank reconciliation showed check No. 748 for $210.90 and check No. 710 for $195.80 outstanding on April 30. The June bank statement included check No. 710 but not check No. 748.

Kinko's checkbook balance		$3,301.82	Bank balance			$2,955.82
Add:			Add:			
Collection on notes receivable	$400.00		Deposit in transit			510.19
Less:						$3,466.01
Collection fee	10.00	390.00	Deduct:			
Error in recording check No. 884		69.00				
		$3,760.82	Outstanding checks:			
			No. 748	$210.90		
			No. 811	110.94		
			No. 912	82.50		404.34
Deduct:						
NSF check	$690.65					
Service charge	8.50	699.15				
Reconciled balance		$3,061.67	Reconciled balance			$3,061.67

END-OF-CHAPTER PROBLEMS

DRILL PROBLEMS (First of Three Sets)

Solve the unknown from the following equations:

5–1.
$$H + 15 = 70$$
$$\underline{-15 \quad\quad -15}$$
$$H = 55$$

5–5.
$$5Y = 75$$
$$\frac{\cancel{5}Y}{\cancel{5}} = \frac{75}{5}$$
$$Y = 15$$

5–9.
$$4(P - 9) = 64$$
$$4P - 36 = 64$$
$$\underline{+36 = +36}$$
$$\frac{\cancel{4}P}{\cancel{4}} = \frac{100}{4}$$
$$P = 25$$

WORD PROBLEMS (First of Three Sets)

5–11. The *Omaha World-Herald,* on January 23, 2001, ran an article titled "Lending Agency Convicted of Predatory Lending Practices." A loan company took the title to an elderly Bellevue widow's home by paying taxes of about $22,200. The market value of the house is $3\frac{1}{2}$ times the tax. What was the market value? Round to the nearest ten thousands.

Unknown(s)	Variable(s)	Relationship
Market value	M	$M = 3\frac{1}{2} \times$ tax

$M = 3\frac{1}{2} \times \$22,200$
$M = \$77,700$
$\quad = \$80,000$

5–13. Joe Sullivan and Hugh Kee sell cars for a Ford dealer. Over the past year, they sold 300 cars. Joe sells 5 times as many cars as Hugh. How many cars did each sell?

Unknown(s)	Variable(s)	Relationship
Hugh	C	C
Joe	$5C$	$+ 5C$
		300 cars

$5C + C = 300$
$\frac{6C}{6} = \frac{300}{6}$
$C = 50$
$C = 50$ (Hugh)
$5C = 250$ (Joe)

5–15. Dots sells T-shirts ($2) and shorts ($4). In April, total sales were $600. People bought 4 times as many T-shirts as shorts. How many T-shirts and shorts did Dots sell? Check your answer.

Unknown(s)	Variable(s)	Price	Relationship
T-shirts	$4S$	$2	$8S$
Shorts	S	$4	$+ 4S$
			600 total sales

$8S + 4S = 600$
$\frac{12S}{12} = \frac{600}{12}$
$S = 50$ shorts
$4S = 200$ T-shirts

Check
$50(\$4) + 200(\$2) = \$600$
$\$200 + \$400 = \$600$
$\$600 = \$600$

DRILL PROBLEMS (Second of Three Sets)

5–17.
$$6B = 420$$
$$\frac{\cancel{6}B}{\cancel{6}} = \frac{420}{6}$$
$$B = 70$$

5–21.
$$9Y - 10 = 53$$
$$\underline{+10 \quad\quad +10}$$
$$\frac{\cancel{9}Y}{\cancel{9}} = \frac{63}{9}$$
$$Y = 7$$

WORD PROBLEMS (Second of Three Sets)

5–23. On a flight from Boston to Los Angeles, American Airlines reduced its Internet price $130. The sale price was $299.50. What was the original price?

Unknown(s)	Variable(s)	Relationship
Original price	P	$P - \$130$ = Sale price
		Sales price = $299.50

$$
\begin{array}{rcl}
P - \$130 & = & \$299.50 \\
+\ 130 & & +\ 130.00 \\
\hline
P & = & \$429.50
\end{array}
$$

5–25. Bill's Roast Beef sells 5 times as many sandwiches as Pete's Deli. The difference between their sales is 360 sandwiches. How many sandwiches did each sell?

Unknown(s)	Variable(s)	Relationship
Bill's	$5S$	$5S$ (450)
Pete's	S	$-\ S$ (90)
		360 sandwiches

$$
\begin{array}{c}
5S - S = 360 \\
\dfrac{4S}{4} = \dfrac{360}{4} \\
S = 90 \\
5S = 450
\end{array}
$$

DRILL PROBLEMS (Third of Three Sets)

Solve the unknown from the following equations.

5–29.
$$
\begin{array}{rcl}
B + 82 - 11 & = & 190 \\
B + 71 & = & 190 \\
-\ 71 & & -\ 71 \\
\hline
B & = & 119
\end{array}
$$

5–31.
$$
\begin{array}{rcl}
3M + 20 & = & 2M + 80 \\
-\ 2M & & -\ 2M \\
\hline
M + 20 & = & +\ 80 \\
-\ 20 & & -\ 20 \\
\hline
M & = & 60
\end{array}
$$

WORD PROBLEMS (Third of Three Sets)

5–33. On December 7, 2000, the *Chicago Sun-Times* ran an article comparing major media ad outlays in the millions of dollars (TV, print, radio, outdoor). In the year 2003, the expected outlay in North America will be $168.6. This is $1\frac{1}{4}$ times more than was spent in 1999. What was the total dollar outlay (in millions) during 1999? Round to nearest tenth.

Unknown(s)	Variable(s)	Relationship
1999 outlay	D	$1\frac{1}{4} D$ = $168.6 million

$$
1\tfrac{1}{4} D = \$168.6 \text{ million}
$$
$$
\frac{1\tfrac{1}{4}D}{1\tfrac{1}{4}} = \frac{\$168.6 \text{ million}}{1\tfrac{1}{4}}
$$
$$
D = \$134.88 \text{ million} = \$134.9 \text{ million}
$$

5–39. Ace Hardware sells cartons of wrenches ($100) and hammers ($300). Howard ordered 40 cartons of wrenches and hammers for $8,400. How many cartons of each are in the order? Check your answer.

Unknown(s)	Variable(s)	Price	Relationship
Wrenches	$40 - H$	$100	$100(40 - H)$
Hammers	H	300	$+\ 300H$
			Total = $8,400

$$
\begin{array}{rcl}
300H + 100(40 - H) & = & 8,400 \\
300H + 4,000 - 100H & = & 8,400 \\
200H + 4,000 & = & 8,400 \\
-\ 4,000 & & -\ 4,000 \\
\hline
\dfrac{200H}{200} & = & \dfrac{4,400}{200}
\end{array}
$$

$H = 22$ cartons of hammers

$40 - H = 18$ cartons of wrenches

Check

$22(\$300) + 18(\$100) = \$8,400$

$\$6,600 + \$1,800 = \$8,400$

$\$8,400 = \$8,400$

CHALLENGE PROBLEM

5–41. Bessy has 6 times as much money as Bob, but when each earns $6, Bessy will have 3 times as much money as Bob. How much does each have before and after earning the $6?

Unknown(s)	Variable(s)	Relationship
Bessy	$6B$	$6B + 6$
Bob	B	$B + 6$

$$6B + 6 = 3(B + 6)$$
$$6B + 6 = 3B + 18$$
$$\underline{-3B \qquad\qquad -3B}$$
$$3B + 6 = 18$$
$$\underline{-6 \qquad\qquad -6}$$
$$\frac{3B}{3} = \frac{12}{3}$$

Before: $B = 4$ After: $B = 10$
$6B = 24$ $6B = 30$

19

END-OF-CHAPTER PROBLEMS

DRILL PROBLEMS

Convert the following decimal to percent:

6–5. 3.561   356.1%

Convert the following percent to decimal:

6–9. $64\frac{3}{10}\%$ .643

Convert the following fraction to percent (round to the nearest tenth percent as needed):

6–13. $\frac{1}{12} = .0833 = 8.3\%$

Convert the following to fractions and reduce to lowest terms:

6–17. 5% $5 \times \frac{1}{100} = \frac{5}{100} = \frac{1}{20}$

6–19. $31\frac{2}{3}\%$ $\frac{95}{3} \times \frac{1}{100} = \frac{95}{300} = \frac{19}{60}$

Solve for the portion (round to the nearest hundredth as needed): $P = R \times B$

6–27. 17.4% of 900
$.174 \times 900 = 156.6$

6–31. 18% of 90
$.18 \times 90 = 16.2$

Solve for the base (round to the nearest hundredth as needed): $\frac{P}{R} = B$

6–33. 170 is 120% of 141.67 $\left(\frac{170}{1.2}\right)$

6–37. 800 is $4\frac{1}{2}\%$ of 17,777.78 $\left(\frac{800}{.045}\right)$

Solve for rate (round to the nearest tenth percent as needed): $\frac{P}{B} = R$

6–41. 110 is 110% of 100 $\left(\frac{110}{100}\right)$

6–43. 16 is 400% of 4 $\left(\frac{16}{4}\right)$

Solve the following problems. Be sure to show your work. Round to the nearest hundredth or hundredth percent as needed:

6–47. 770 is 70% of what number? $\frac{770}{.7} = 1,100$ $\frac{P}{R} = B$

6–49. What percent of 150 is 60? $\frac{60}{150} = 40\%$ $\frac{P}{B} = R$

Complete the following table:

| Product | Sales in millions | | Amount of decrease or increase | Percent change (to nearest hundredth percent as needed) |
	2004	2005		
6–51. DVD players	$ 50	$ 47	− $ 3	$-6\% \left(\frac{\$3}{\$50}\right)$

WORD PROBLEMS (First of Four Sets)

6–53. What percent of customers in Problem 6–52 did not order Diet Coke?
Excel $\frac{4,500}{6,000} = 75\%$

Note: Portion and rate must refer to same piece of the base.

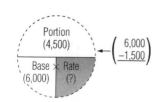

6–57. Pete Lavoie went to JCPenney and bought a Sony CD player. The purchase price was $350. He made a down payment of 30%. How much was Pete's down payment?

.30 × $350 = $105

6–61. Christie's Auction sold a painting for $24,500. It charges all buyers a 15% premium of the final bid price. How much did the bidder pay Christie's?

$24,500 × 1.15 = $28,175   *Note:* Portion is larger than base since rate is greater than 100%.

WORD PROBLEMS (Second of Four Sets)

6–63. What percent of college students in Problem 6–62 eat breakfast?

$$\frac{5,400}{6,000} = .90 = 90\%$$

Note: Portion and rate refer to same part of base.

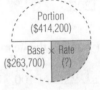

6–65. On May 30, 2001, Associated Press Online provided information from a fiscal analyst stating that during the past 20 years, the average after-tax income of the wealthiest 1% of Americans had grown from $263,700 to $677,900. What was the percent increase? Round to the nearest percent.

$$\begin{array}{r} \$677,900 \\ -\ 263,700 \\ \hline \$414,200 \text{ increase} \end{array}$$

$$\frac{\$414,200}{\$263,700} = 157.07243\%$$

$$= 157\%$$

6–71. Blue Valley College has 600 female students. This is 60% of the total student body. How many students attend Blue Valley College?

$$\frac{600}{.60} = 1,000$$

6–75. Borders bookstore ordered 80 marketing books but received 60 books. What percent of the order was missing?

$$\frac{20}{80} = 25\%$$

WORD PROBLEMS (Third of Four Sets)

6–77. Referring to increased mailing costs, the May 14, 2001, *Advertising Age* commented "Magazines Go Postal Over Rate Hike." The new rate will cost publishers $50 million; this is 12.5% more than they paid the previous year. How much did it cost publishers last year? Round to the nearest hundreds.

$$\frac{\$50{,}000{,}000}{1.125} = \$44{,}444{,}444$$
$$= \$44{,}444{,}400$$

6–81. Web Consultants, Inc., pays Alice Rose an annual salary of $48,000. Today, Alice's boss informs her that she will receive a $6,400 raise. What percent of Alice's old salary is the $6,400 raise? Round to nearest tenth percent.

$$\frac{\$6{,}400}{\$48{,}000} = 13.3\%$$

6–83. Petco ordered 100 dog calendars but received 60. What percent of the order was missing?

$$\frac{40}{100} = 40\%$$

WORD PROBLEMS (Fourth of Four Sets)

6–89. John O'Sullivan has just completed his first year in business. His records show that he spent the following in advertising:

Newspaper	$600	Radio	$650
Yellow Pages	700	Local flyers	400

What percent of John's advertising was spent on the Yellow Pages? Round to the nearest hundredth percent.
$600 + $700 + $650 + $400 = $2,350

$$\frac{\$700}{\$2{,}350} = 29.79\%$$

6–93. On May 25, 2001, the *Chicago Sun-Times* had an article titled "What's Its Worth at Resale?" The article compared the resale benefit of various remodeling costs. Refinishing your basement at a cost of $45,404 would add $18,270 to the resale value of your home. What percent of your cost is recouped? Round to the nearest percent.

$$\frac{\$18{,}270}{\$45{,}404} = 40.238745\%$$
$$= 40\%$$

6–95. Assume 450,000 people line up on the streets to see the Macy's Thanksgiving Parade in 2003. If attendance is expected to increase 30%, what will be the number of people lined up on the street to see the 2004 parade?

$450,000 \times 1.30 = 585,000$

CHALLENGE PROBLEM

6–97. A local Dunkin' Donuts shop reported that its sales have increased exactly 22% per year for the last 2 years. This year's sales were $82,500. What were Dunkin' Donuts sales 2 years ago? Round each year's sales to the nearest dollar.

$\dfrac{\$82,500}{1.22} = \$67,623$ sales last year $\dfrac{\$67,623}{1.22} = \$55,429$

END-OF-CHAPTER PROBLEMS

DRILL PROBLEMS

For all problems, round your final answer to the nearest cent. Do not round net price equivalent rates or single equivalent discount rates.

Complete the following:

Item	List price	Chain discount	Net price equivalent rate (in decimals)	Single equivalent discount rate (in decimals)	Trade discount	Net price
7–1. Sony Playstation	$399	7/2	.9114	.0886	$35.35	$363.65

$$
\begin{array}{cc}
1.00 & 1.00 \\
-\ .07 & -\ .02 \\
\hline
.93\ \times & .98
\end{array} = .9114 \times \$399 = \$363.65
$$

$$
\begin{array}{c}
1.0000 \\
-\ .9114 \\
\hline
.0886
\end{array} \times \$399 = \$35.35
$$

Complete the following:

Item	List price	Chain discount	Net price	Trade discount
7–5. Maytag dishwasher	$450	8/5/6	$369.70	$80.30

$450 \times .82156 = \$369.70\ (.82156 = .92 \times .95 \times .94)$
$450 \times .17844 = \$80.30\ (1 - .82156 = .17844)$

7–7. Land Rover roofrack	$1,850	12/9/6	$1,392.59	$457.41

$\$1,850 \times .752752 = \$1,392.59\ (.752752 = .88 \times .91 \times .94)$
$\$1,850 \times .247248 = \$457.41\ (1 - .752752 = .247248)$

Complete the following:

Invoice	Dates when goods received	Terms	Last day* of discount period	Final day bill is due (end of credit period)
7–9. June 18		1/10, n/30	June 28	July 18

By Table 7.1, June 18 = 169 + 30 = 199 ——————→ Search in Table 7.1.

7–13. June 12		3/10 EOM	July 10	July 30 Discount and credit period begin at end of month of sale.

Complete the following by calculating the cash discount and net amount paid:

Gross amount of invoice (freight charge already included)	Freight charge	Date of invoice	Terms of invoice	Date of payment	Cash discount	Net amount paid
7–15. $7,000	$100	4/8	2/10, n/60	4/15	$138 (.02 × $6,900)	$6,862 ($6,900 × .98 = $6,762 + $100 = $6,862)

Complete the following:

	Amount of invoice	Terms	Invoice date	Actual partial payment made	Date of partial payment	Amount of payment to be credited	Balance outstanding
7–19.	$700	2/10, n/60	5/6	$400	5/15	$408.16	$291.84

$$\frac{\$400}{.98} = \begin{array}{r} \$700.00 \\ -\ 408.16 \\ \hline \$291.84 \end{array}$$

WORD PROBLEMS (Round to Nearest Cent as Needed)

7–21. The list price of a Fossil watch is $120.95. Jim O'Sullivan receives a trade discount of 40%. Find the trade discount amount and the net price.
$120.95 \times .40 = \$48.38$
$120.95 \times .60 = \$72.57$

7–25. A manufacturer of skateboards offered a 5/2/1 chain discount to many customers. Bob's Sporting Goods ordered 20 skateboards for a total $625 list price. What was the net price of the skateboards? What was the trade discount amount?

Net price	Trade discount
$.95 \times .98 \times .99 = .92169 \times \$625 = \$576.05625$	$.07831 \times \$625 = \$48.94$
$= \$576.06$	

7–29. Macy of New York sold Marriott of Chicago office equipment with a $6,000 list price. Sale terms were 3/10, n/30 FOB New York. Macy agreed to prepay the $30 freight. Marriott pays the invoice within the discount period. What does Marriott pay Macy?
$.97 \times \$6,000 = \$5,820 + \$30 \text{ freight} = \$5,850$

7–33. On August 1, Intel Corporation (Problem 7–32) returns $100 of the machinery due to defects. What does Intel pay Bally on August 5? Round to nearest cent.

$$\begin{array}{r} \$14,000 \\ -\ \ \ \ 100 \ \text{returns} \\ \hline \$13,900 \end{array} \times .60 = \$8,340 \qquad \$8,340 \times .98 = \$8,173.20$$

ADDITIONAL SET OF WORD PROBLEMS

7–39. Vail Ski Shop received a $1,201 invoice dated July 8 with 2/10, 1/15, n/60 terms. On July 22, Vail sent a $485 partial payment. What credit should Vail receive? What is Vail's outstanding balance?

$$\frac{\$485}{.99} = \$489.90 \qquad \begin{array}{r} \$1,201.00 \\ -\ 489.90 \\ \hline \$\ \ 711.10 \ \text{balance outstanding} \end{array}$$

7–41. Staples purchased seven new computers for $850 each. It received a 15% discount because it purchased more than five and an additional 6% discount because it took immediate delivery. Terms of payment were 2/10, n/30. Staples pays the bill within the cash discount period. How much should the check be? Round to the nearest cent.

$$\begin{array}{r} \$\ \ 850.00 \\ \times\ \ \ \ \ \ 7 \\ \hline \$5,950.00 \end{array} \qquad \begin{array}{r} \$5,950.00 \\ \times\ .799 \ \text{chain discount} \\ \hline \$4,754.05 \\ \times\ .98 \ \text{cash discount} \\ \hline \$4,658.97 \end{array}$$

Chain: $.85 \times .94 = .799$

7–43. Sam's Ski Boards.com offers 5/4/1 chain discounts to many of its customers. The Ski Hut ordered 20 ski boards with a total list price of $1,200. What is the net price of the ski boards? What was the trade discount amount? Round to the nearest cent.

.95 × .96 × .99 = .90288

$ 1,200
× .90288
$1,083.46 net price

$ 1,200
× .09712 (1 − .90288)
$116.54 trade discount

CHALLENGE PROBLEM

7–49. In the October 9, 2000, issue of *Crain's New York Business,* a story appeared describing firms saving money by taking a plunge into online buying pools. Karen Curry was eager to find seven dozen soft-sided drink coolers for a company promotion. She found a good price, $24 each. However, through Mercata.com, a group purchasing website that offers volume discounts to smaller companies who electronically pool their orders, she found a lower price of $21. **(a)** What was Karen's final total net purchase price? **(b)** What was her total discount amount using the website? **(c)** What was the percent discount using the online buying pool? **(d)** If Karen meets the cash discount period of 2/10 net 30, what would be her final price?

a. $21 × 84 (7 dozen) = $1,764 paid using the Web

b. $24 × 84 = $2,016 not using the Web
− 1,764
$ 252 savings (discount amount)

c. $\dfrac{\$252 \,(P)}{\$2,016 \,(B)} = 12.5\%$

d. $1,764 × .98 = $1,728.72

END-OF-CHAPTER PROBLEMS

DRILL PROBLEMS

Assume markups in Problems 8–1 to 8–6 are based on cost. Find the dollar markup and selling price for the following problems. Round answers to the nearest cent.

	Item	Cost	Markup percent	Dollar markup	Selling price
8–1.	Sony DVD player	$100	40%	$40	$140

$S = C + M$

$S = \$100 + .40(\$100)$ **Check** $S = \text{Cost} \times (1 + \text{Percent markup on cost})$

$S = \$100 + \$40$

$S = \$140$ $\$140 = \$100 \times 1.40$

Solve for cost (round to the nearest cent):

8–3. Selling price of office furniture at Staples, $6,000 $\$6,000 = C + .40C$

Percent markup on cost, 40% $\dfrac{\$6,000}{1.40} = \dfrac{1.40C}{1.40}$

Actual cost? **Check** $\dfrac{\$6,000}{1.40} = \$4,285.71$ $\$4,285.71 = C$ $C = \dfrac{\text{Selling price}}{1 + \text{Percent markup on cost}}$

Complete the following:

	Cost	Selling price	Dollar markup	Percent markup on cost*
8–5.	$15.10	$22.00	? $6.90	? $45.70\% \left(\dfrac{\$6.90}{\$15.10} \right)$
8–6.	? $4.60	? $9.30	$4.70	102.17% $C = \dfrac{\$4.70}{1.0217}$

*Round to the nearest hundredth percent.

Assume markups in Problems 8–7 to 8–12 are based on selling price.
Find the dollar markup and cost (round answers to the nearest cent):

	Item	Selling price	Markup percent	Dollar markup	Cost
8–7.	Kodak digital camera	$219	30%	$65.70	$153.30

$\$219.00 = C + .30(\$219)$ **Check**

$\$219.00 = C + \$65.70$ *Note:* Markup $\$153.30 = \$219 \times .70$

$\underline{-\ 65.70} \quad \underline{-\ 65.70}$ is on selling

$\$153.30 = C$ price, not cost. $C = \text{Selling price} \times (1 - \text{Percent markup on selling price})$

Solve for the selling price (round to the nearest cent):

8–9. Selling price of a complete set of pots and pans at Wal-Mart?

40% markup on selling price $S = \$66.50 + .40S$

Cost, actual, $66.50 $\underline{-\ .40S} \qquad \underline{-\ .40S}$

Check $\dfrac{\$66.50}{.60} = \$110.83$ $\dfrac{.60S}{.60} = \dfrac{\$66.50}{.60}$

$S = \$110.83$

$S = \dfrac{\text{Cost}}{1 - \text{Percent markup on selling price}}$

Complete the following:

	Cost	Selling price	Dollar markup	Percent markup on selling price (round to nearest tenth percent)
8–11.	$14.80	$49.00	? $34.20	? 69.8% $\left(\dfrac{\$34.20}{\$49.00}\right)$

Complete the following:

8–15. Calculate the final selling price to the nearest cent and markdown percent to the nearest hundredth percent:

Excel	Original selling price	First markdown	Second markdown	Markup	Final markdown
	$5,000	20%	10%	12%	5%

$5,000 × .80 = \$4,000.00$

$4,000 × .90 = \$3,600.00$

$3,600 × 1.12 = \$4,032.00$

$4,032 × .95 = \$3,830.40$

$$\begin{array}{r} \$5,000.00 \\ -\ 3,830.40 \\ \hline \$1,169.60 \end{array}$$

$\dfrac{\$1,169.60}{\$5,000.00} = 23.39\%$

WORD PROBLEMS

8–17. On an eBay auction, Mike Kaminsky bought an old Walter Lantz Woody Woodpecker oil painting for $4,000. He plans to resell it at a toy show for $7,000. What are the dollar markup and the percent markup on cost? Check the cost figure.

Dollar markup $= S - C$

$\$3,000 = \$7,000 - \$4,000$

Percent markup on cost $= \dfrac{\$3,000}{\$4,000} = 75\%$

Check $C = \dfrac{\text{Dollar markup}}{\text{Percent markup on cost}} = \dfrac{\$3,000}{.75} = \$4,000$

8–23. Misu Sheet, owner of the Bedspread Shop, knows his customers will pay no more than $120 for a comforter. Misu wants a 30% markup on selling price. What is the most that Misu can pay for a comforter?

Note: The markup is on the selling price, not cost.

$\$120 = C + .30(\$120)$

$\$120 = C + \$36$

$$\begin{array}{r} -\ 36 \qquad -\ 36 \\ \hline \$\ 84 = C \end{array}$$

Check

$C = \text{Selling price} \times (1 - \text{Percent markup on selling price})$

$\$84 = \$120 \times .70$

8–24. Assume Misu Sheet (Problem 8–23) wants a 30% markup on cost instead of on selling price. What is Misu's cost? Round to the nearest cent.

Note: The markup is on the cost, not the selling price.

$\$120 = C + .30C$

$\dfrac{\$120}{1.3} = \dfrac{1.3C}{1.3}$

$\$92.31 = C$

Check

$C = \dfrac{\text{Selling price}}{1 + \text{Percent markup on cost}}$

$\$92.31 = \dfrac{\$120}{1.30}$

ADDITIONAL SET OF WORD PROBLEMS

8–29. Sears bought a treadmill for $510. Sears has a 60% markup on selling price. What is the selling price of the treadmill?

Note: Markup is on selling price, not cost.

$S = \$510 + .60S$

$$\begin{array}{r} -\ .60S \qquad -\ .60S \\ \hline \dfrac{.40S}{.40} = \dfrac{\$510}{.40} \end{array}$$

$S = \$1,275$

Check $S = \dfrac{\text{Cost}}{1 - \text{Percent markup on selling price}}$

$\$1,275 = \dfrac{\$510}{.40}$

8–35. Arley's Bakery makes fat-free cookies that cost $1.50 each. Arley expects 15% of the cookies to fall apart and be discarded. Arley wants a 45% markup on cost and produces 200 cookies. What should Arley price each cookie? Round to the nearest cent.

Total cost = $200 \times \$1.50 = \$300$

$TS = \$300 + .45(\$300)$

$TS = \$300 + \$135$

$TS = \$435$

$$\frac{\$435}{170 \text{ cookies}} = \$2.56$$

CHALLENGE PROBLEM

8–39. On July 8, 2004, Leon's Kitchen Hut bought a set of pots with a $120 list price from Lambert Manufacturing. Leon's receives a 25% trade discount. Terms of the sale were 2/10, n/30. On July 14, Leon's sent a check to Lambert for the pots. Leon's expenses are 20% of the selling price. Leon's must also make a profit of 15% of the selling price. A competitor marked down the same set of pots 30%. Assume Leon's reduces its selling price by 30%.

a. What is the sale price at Kitchen Hut?

b. What was the operating profit or loss?

a. $\$120 \times .75 = \$90 \times .98 = \$88.20$ **b.** Total cost $= \$88.20 + .20(\$135.69)$ $P = SP - TC$

$\quad S = \$88.20 + .20S + .15S$ $= \$88.20 + \$27.14$ $= \$94.98 - \$115.34$

$\quad S = \$88.20 + .35S$ $= \$115.34$ Loss = \$20.36

$.65S = \$88.20$

$\quad S = \$135.69$

Sale price: $94.98, or (.70 $\times$ $135.69)

31

1. Assume Kellogg's produced 715,000 boxes of Corn Flakes this year. This was 110% of the annual production last year. What was last year's annual production? (p. 142)

$$\frac{715,000}{1.10} = 650,000$$

5. Runners World marks up its Nike jogging shoes 25% on selling price. The Nike shoe sells for $65. How much did the store pay for them? (p. 201)

$$S = C + M$$
$$\$65.00 = C + .25(\$65)$$
$$\$65.00 = C + \$16.25$$
$$\underline{-16.25 \qquad\quad -16.25}$$
$$\$48.75 = C$$

$C =$ Selling price $\times (1 -$ Percent markup on selling price$)$

$$C = \$65 \times .75$$
$$C = \$48.75$$

7. Bonnie's Bakery bakes 60 loaves of bread for $1.10 each. Bonnie's estimates that 10% of the bread will spoil. Assume a 60% markup on cost. What is the selling price of each loaf? If Bonnie's can sell the old bread for one-half the cost, what is the selling price of each loaf? (p. 206)

$$TC = 60 \times \$1.10 = \$66$$
$$TS = TC + TM$$
$$TS = \$66 + .60(\$66)$$
$$TS = \$66 + \$39.60$$
$$TS = \$105.60$$

$$\frac{\$105.60}{54} = \$1.96$$
$$(60 \times .90)$$

$$\frac{\$105.60 - (6 \times \$.55)}{54} = \frac{\$105.60 - \$3.30}{54} = \$1.89$$

END-OF-CHAPTER PROBLEMS

DRILL PROBLEMS

Complete the following table (assume the overtime for each employee is a time-and-a-half rate after 40 hours):

Employee	Gross earnings	M	T	W	Th	F	Sa	Total regular hours	Total overtime hours	Regular rate	Overtime rate
9–3. Blue	$452.00	12	9	9	9	9	3	40	11	$8.00	$12.00

$40 \times \$8 = \$320$
$11 \times \$12 = \underline{\hphantom{0}132}$
$\$452$

Calculate gross earnings:

Worker	Number of units produced	Rate per unit	Gross earnings
9–5. Lang	510	$2.10	$1,071.00 (510 × $2.10)

Calculate the gross earnings for each apple picker based on the following differential pay scale:

1–1,000:	$.03 each	1,001–1,600	$.05 each	over 1,600	$.07 each

Apple picker	Number of apples picked	Gross earnings
9–7. Ryan	1,600	$60 = (1,000 × $.03) + (600 × $.05)

Ron Company has the following commission schedule:

Commission rate	Sales
2%	Up to $80,000
3.5%	Excess of $80,000 to $100,000
4%	More than $100,000

Calculate the gross earnings of Ron Company's two employees:

Employee	Total sales	Gross earnings
9–10. Bill Moore	$ 70,000	$1,400 ($70,000 × .02)
9–11. Ron Ear	$155,000	

$$\begin{pmatrix} \$80,000 \times .02 = \$1,600 \\ \$20,000 \times .035 = 700 \\ \$55,000 \times .04 = \underline{2,200} \\ \$4,500 \end{pmatrix}$$

Complete the following payroll register. Calculate FIT by the percentage method for this weekly period; Social Security and Medicare are the same rates as in the previous problems. No one will reach the maximum for FICA.

					FICA		
Employee	Marital status	Allowances claimed	Gross pay	FIT	S.S.	Med.	Net pay
9–17. Al Holland	M	2	$1,200	$161.37	$74.40	$17.40	$946.83

S.S.: $1,200 × .062 = $74.40   Med.: $1,200 × .0145 = $17.40

$$
\begin{array}{rl}
\text{FIT:} & \$1,200.00 \\
- & 111.54 \; (\$55.77 \times 2) \\
\hline
& \$1,088.46 \\
- & 960.00 \\
\hline
& \$\;\;128.46
\end{array}
$$

$$
\begin{array}{l}
\$125.40 \\
+ \;\; 35.97 \; (.28 \times \$128.46) \\
\hline
\$161.37
\end{array}
$$

9–19. Given the following, calculate the state (assume 5.3%) and federal unemployment taxes that the employer must pay for each of the first two quarters. The federal unemployment tax is .8% on the first $7,000.

Payroll summary		
	Quarter 1	Quarter 2
Bill Adams	$4,000	$ 8,000
Rich Haines	8,000	14,000
Alice Smooth	3,200	3,800

*Note only first $7,000 is taxed.

†Only first $3,000 is taxed since that puts Adams over $7,000 for the year.

Quarter 1

Adams	$ 4,000
Haines	7,000*
Smooth	3,200
	$14,200
	× .053
SUTA =	$752.60

Quarter 2

Adams	$ 3,000†
Haines	0
Smooth	3,800
	$ 6,800
	× .053
SUTA =	$360.40

FUTA = $ 54.40 ($6,800 × .008)

.008 × $14,200 = $113.60 FUTA

WORD PROBLEMS

9–21. The March 13, 2001, issue of the *Chicago Tribune* reported that stagehands would be paid $22.40 per hour. Bill Crew, a stagehand, worked 33 hours. **(a)** How much will be deducted from Bill's pay for Social Security? **(b)** How much will be deducted for Medicare?

a.
$$
\begin{array}{rl}
\$ \;\; 22.40 & \text{per hour} \\
\times \;\;\;\; 33 & \text{hours worked} \\
\hline
\$ \; 739.20 & \text{total wages} \\
\times \;\;\; .062 & \text{(Social Security tax)} \\
\hline
\$45.8304 & = \$45.83 \text{ Social Security}
\end{array}
$$

b.
$$
\begin{array}{rl}
\$ \; 739.20 & \\
\times \;\; .0145 & \text{(Medicare tax)} \\
\hline
\$10.7184 & = \$10.72
\end{array}
$$

9–23. Dennis Toby is a salesclerk at Northwest Department Store. Dennis receives $8 per hour plus a commission of 3% on all sales. Assume Dennis works 30 hours and has sales of $1,900. What is his gross pay?

(30 hours × $8) + ($1,900 × .03) = $240 + $57 = $297

9–27. Richard Gaziano is a manager for Health Care, Inc. Health Care deducts Social Security, Medicare, and FIT (by percentage method) from his earnings. Assume the same Social Security and Medicare rates as in Problem 9–26. Before this payroll, Richard is $1,000 below the maximum level for Social Security earnings. Richard is married, is paid weekly, and claims 2 exemptions. What is Richard's net pay for the week if he earns $1,300?

Social Security: $1,000 × .062 = $62   Medicare: $1,300 × .0145 = $18.85

$1,300 − $62 − $18.85 − $189.37 = $1,029.78

$$
\begin{array}{rl}
\text{FIT:} & \$1,300.00 \\
- & 111.54 \; (\$55.77 \times 2) \\
\hline
& \$1,188.46 \\
- & 960.00 \\
\hline
& \$\;\;228.46
\end{array}
$$

$125.40 + .28($228.46)

$125.40 + $63.97 = $189.37

34

CHALLENGE PROBLEM

9–31. On January 14, 2001, the *Miami Herald* reported that a school psychologist earns $50,000 a year in the Miami-Dade public school system. She is single and claims one exemption. **(a)** What would be taken out of her check each month for Social Security? **(b)** What would be taken out each month for Medicare? **(c)** What would be taken out for FIT using the percentage method?

a. $\dfrac{\$50,000 \text{ per year}}{12} = \$4,166.666 = \$4,166.67$ earned per month

$$
\begin{array}{r}
\$\ 4,166.67 \\
\times \quad\ .062 \quad \text{Social Security} \\
\hline
\$258.33354 = \$258.33 \text{ Social Security tax}
\end{array}
$$

b.
$$
\begin{array}{r}
\$\ 4,166.67 \\
\times \quad\ .0145 \\
\hline
\$60.416715 = \$60.42 \text{ Medicare tax}
\end{array}
$$

c. FIT:

$$
\begin{array}{r}
\$241.67 \text{ monthly pay} \\
\times \quad\ 1 \text{ exemption} \\
\hline
\$241.67
\end{array}
$$

$$
\begin{array}{r}
\$4,166.67 \\
- \quad 241.67 \\
\hline
\$3,925.00 \\
- \ 2,392.00 \\
\hline
\$1,533.00 \ \times \ .28 =
\end{array}
\quad
\begin{array}{r}
\$325.65 \\
+ \ 429.24 \\
\hline
\$754.89 \ \text{FIT}
\end{array}
$$

END-OF-CHAPTER PROBLEMS

DRILL PROBLEMS

Calculate the simple interest and maturity value for the following problems. Round to the nearest cent as needed.

	Principal	Interest rate	Time	Simple interest	Maturity value
10–3.	$7,000	$8\frac{1}{4}\%$	7 mo.	$336.88	$7,336.88

$$\$7,000 \times .0825 \times \frac{7}{12} = \$336.88$$

Complete the following, using ordinary interest: $T = \dfrac{\text{Exact no. of days}}{360}$

	Principal						
10–5.	$585	9%	June 5	Dec. 15	193	$28.23	$613.23
Excel			156	349			

$$\$585 \times .09 \times \frac{193}{360} = \$28.23$$

Complete the following, using exact interest: $T = \dfrac{\text{Exact no. of days}}{365}$

	Principal	Interest rate	Date borrowed	Date repaid	Exact time	Interest	Maturity value
10–7.	$1,000	8%	Mar. 8	June 9	93	$20.38	$1,020.38
			67	160			

$$\$1,000 \times .08 \times \frac{93}{365} = \$20.38$$

10–13. Use the U.S. Rule to solve for total interest costs, balances, and final payments (use ordinary interest).

Given Principal: $10,000, 8%, 240 days
Partial payments: On 100th day, $4,000
On 180th day, $2,000

8%, 100 days, $10,000

$$I = \$10,000 \times .08 \times \frac{100}{360} = \$222.22$$

$$\begin{array}{r} \$4,000.00 \\ -\quad 222.22 \\ \hline \$3,777.78 \end{array}$$

$$\begin{array}{r} \$10,000.00 \\ -\quad 3,777.78 \\ \hline \$\ 6,222.22 \ \text{adjusted balance} \end{array}$$

$$\$6,222.22 \times .08 \times \frac{80}{360} = \$110.62$$

$$\begin{array}{r} \$2,000.00 \\ -\quad 110.62 \\ \hline \$1,889.38 \end{array}$$

$$\begin{array}{r} \$6,222.22 \\ -\quad 1,889.38 \\ \hline \$4,332.84 \ \text{adjusted balance} \end{array}$$

$$\$4,332.84 \times .08 \times \frac{60}{360} = \$57.77$$

$$\begin{array}{r} \$4,332.84 \\ +\quad 57.77 \\ \hline \$4,390.61 \ \text{balance due} \end{array}$$

Interest paid

$$\begin{array}{r} \$222.22 \\ 110.62 \\ +\quad 57.77 \\ \hline \$390.61 \end{array}$$

WORD PROBLEMS

10–17. Kelly O'Brien met Jody Jansen (Problem 10–16) at Sunshine Bank and suggested she consider the loan on exact interest. Recalculate the loan for Jody under this assumption.

$$\$2,300 \times .09 \times \frac{137}{365} = \$77.70 + \$2,300 = \$2,377.70 \qquad \text{Save } \$1.08$$

10–19. Gordon Rosel went to his bank to find out how long it will take for $1,200 to amount to $1,650 at 8% simple interest. Please solve Gordon's problem. Round time in years to the nearest tenth.

$$\frac{\$450}{\$1,200 \times .08} = \frac{\$450}{\$96} = 4.7 \text{ years}$$

ADDITIONAL SET OF WORD PROBLEMS

10–25. The January 2001 issue of *Consumer Reports* gave a report on interest charges. The article mentioned that with a good payment history you could be paying only 9% to 12% on credit cards. With a $560 charge, what would be the interest amount at 9% and at 12% for 1 month?

$$\overset{P}{\$560} \times \overset{R}{.09} \times \overset{T}{\frac{1}{12}} = \$4.20 \text{ interest}$$

$$\overset{P}{\$560} \times \overset{R}{.12} \times \overset{T}{\frac{1}{12}} = \$5.60 \text{ interest}$$

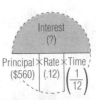

10–29. Margie Pagano is buying a car. Her June monthly interest at $12\frac{1}{2}\%$ was $195. What was Margie's principal balance at the beginning of June? Use 360 days. Do not round the denominator before dividing.

$$\frac{\$195}{.125 \times \frac{30}{360}} = \frac{\$195}{.0104166} = \$18,720.12$$

10–31. Carol Miller went to Europe and forgot to pay her $740 mortgage payment on her New Hampshire ski house. For her 59 days overdue on her payment, the bank charged her a penalty of $15. What was the rate of interest charged by the bank? Round to the nearest hundredth percent (assume 360 days).

$$R = \frac{\$15}{\$740 \times \frac{59}{360}} = 12.37\%$$

CHALLENGE PROBLEM

10–37. Janet Foster bought a computer and printer at Computerland. The printer had a $600 list price with a $100 trade discount and 2/10, n/30 terms. The computer had a $1,600 list price with a 25% trade discount but no cash discount. On the computer, Computerland offered Janet the choice of (1) paying $50 per month for 17 months with the 18th payment paying the remainder of the balance or (2) paying 8% interest for 18 months in equal payments.

a. Assume Janet could borrow the money for the printer at 8% to take advantage of the cash discount. How much would Janet save (assume 360 days)?

b. On the computer, what is the difference in the final payment between choices 1 and 2?

a. $\$490 \times .08 \times \dfrac{20}{360} = \$2.18$

($600 − $100) × .98
↑
$10.00
− 2.18
$ 7.82 (savings—worth borrowing)

b. (1) $50 × 17 = $850   Last payment $1,200 − $850 = $350

(2) $1,200 × .08 × 1.5 = $144

$1,200 + $144 = $\dfrac{\$1,344}{18}$ = $74.67

$350.00
− 74.67
$275.33

END-OF-CHAPTER PROBLEMS

DRILL PROBLEMS

Complete the following table for these simple discount notes. Use the ordinary interest method.

	Amount due at maturity	Discount rate	Time	Bank discount	Proceeds
11–1.	$18,000	$4\frac{3}{4}\%$	190 days	$451.25	$17,548.75

$$\$18,000 \times .0475 \times \frac{190}{360} = \$451.25 \qquad \$18,000 - \$451.25 = \$17,548.75$$

Calculate the discount period for the bank to wait to receive its money:

	Date of note	Length of note	Date note discounted	Discount period
11–3.	April 12	45 days	May 2	$45 - 20 = 25$

$$\begin{array}{ll} \text{May 2} & 122 \text{ days} \\ \text{Apr. 12} & -\ 102 \\ \hline & 20 \text{ days} \end{array}$$

Solve for maturity value, discount period, bank discount, and proceeds (assume for Problems 11–5 and 11–6 a bank discount rate of 9%).

	Face value (principal)	Rate of interest	Length of note	Maturity value	Date of note	Date note discounted	Discount period	Bank discount	Proceeds
11–5.	$50,000	11%	95 days	$51,451.39	June 10	July 18	57	$733.18	$50,718.21

$$\begin{array}{ll} \text{July 18} & 199 \text{ days} \\ \text{June 10} & -\ 161 \\ \hline & 38 \text{ days} \end{array}$$

$$\$50,000 \times .11 \times \frac{95}{360} = \$1,451.39 + \$50,000 = \$51,451.39 \ MV$$

Discount period $= 95 - 38 = 57$

Bank discount $= \$51,451.39 \times .09 \times \frac{57}{360} = \$733.18$

Proceeds $= \$51,451.39 - \$733.18 = \$50,718.21$

WORD PROBLEMS

Use ordinary interest as needed.

11–9. Jack Tripper signed a $9,000 note at Fleet Bank. Fleet charges a $9\frac{1}{4}\%$ discount rate. If the loan is for 200 days, find **(a)** the proceeds and **(b)** the effective rate charged by the bank (to the nearest tenth percent).

a. $\$9,000 \times .0925 \times \frac{200}{360} = \$462.50$

$\$9,000 - \$462.50 = \$8,537.50$

b. $\dfrac{\$462.50}{\$8,537.50 \times \frac{200}{360}} = \dfrac{\$462.50}{\$4,743.0555} = 9.8\%$

11–11. On September 5, Sheffield Company discounted at Sunshine Bank a $9,000 (maturity value), 120-day note dated June 5. Sunshine's discount rate was 9%. What proceeds did Sheffield Company receive?

$$\begin{array}{ll} \text{Sept. 5} & 248 \text{ days} \\ \text{June 5} & -\ 156 \\ \hline & 92 \text{ days passed} \end{array}$$

$120 - 92 = 28$ days
            (discount period)

$\$9,000 \times .09 \times \frac{28}{360} = \$63 \qquad \$9,000 - \$63 = \$8,937$

11–17. Hafers, an electrical supply company, sold $4,800 of equipment to Jim Coates Wiring, Inc. Coates signed a promissory note May 12 with 4.5% interest. The due date was August 10. Short of funds, Hafers contacted Charter One Bank on July 20; the bank agreed to take over the note at a 6.2% discount. What proceeds will Hafers receive?

Aug. 10	222 days
May 12	− 132 days
	90 length of loan

$$\$4,800 \times .045 \times \frac{90}{360} = \$54.00$$

$$\$4,800 + \$54.00 = \$4,854.00 \; MV$$

July 20	201 days
May 12	− 132 days
	69 days passed

$$\$4,854.00 \times .062 \times \frac{21}{360} = \$17.56 \qquad \textbf{Bank discount}$$

$$90 - 69 = 21 \text{ days}$$
$$\text{(discount period)}$$

$$\begin{array}{r} \$4,854.00 \\ -\quad 17.56 \\ \hline \$4,836.44 \text{ proceeds} \end{array}$$

CHALLENGE PROBLEM

11–19. Tina Mier must pay a $2,000 furniture bill. A finance company will loan Tina $2,000 for 8 months at a 9% discount rate. The finance company told Tina that if she wants to receive exactly $2,000, she must borrow more than $2,000. The finance company gave Tina the following formula:

$$\text{What to ask for} = \frac{\text{Amount in cash to be received}}{1 - (\text{Discount rate} \times \text{Time of loan})}$$

Calculate Tina's loan request and the effective rate of interest to nearest hundredth percent.

$$\frac{\$2,000}{1 - \left(.09 \times \dfrac{8}{12}\right)} = \frac{\$2,000}{1 - .06} = \frac{\$2,000}{.94} = \$2,127.66$$

Check

$$\$2,127.66 \times .09 \times \frac{8}{12} = \$127.66$$

$$\frac{\$127.66}{\$2,000 \times \dfrac{8}{12}} = 9.57\%$$

42

END-OF-CHAPTER PROBLEMS

DRILL PROBLEMS

Complete the following without using Table 12.1 (round to the nearest cent for each calculation) and then check by Table 12.1 (check will be off one cent due to rounding).

	Principal	Time (years)	Rate of compound interest	Compounded	Periods	Rate	Total amount	Total interest
12–1.	$700	2	6%	Semiannually	4	3%	$787.86	$87.86

2 years × 2 = 4 periods $\dfrac{6\%}{2} = 3\%$

$700.00	$721.00	$742.63	$764.91	$787.86		
× 1.03	× 1.03	× 1.03	× 1.03	− 700.00	**Check**	$700 × 1.1255 = $787.85
$721.00	$742.63	$764.91	$787.86	$ 87.86		

Complete the following using compound future value Table 12.1:

	Time	Principal	Rate	Compounded	Amount	Interest
12–3.	6 months	$10,000	8%	Quarterly	$10,404.00	$404.00

$\dfrac{6}{12} \times 4 = 2$ periods $\dfrac{8\%}{4} = 2\%$

$10,000 × 1.0404 = $10,404

$$
\begin{array}{r}
\$10,404 \\
-\ 10,000 \\
\hline
\$\ \ \ 404
\end{array}
$$

Calculate the effective rate (APY) of interest for 1 year.

12–5. Principal: $15,500 $15,500 × 1.1255 = $17,445.25 $\dfrac{\$1,945.25}{\$15,500} = .1255 = 12.55\%$

Interest rate: 12%   − 15,500.00

Compounded quarterly $ 1,945.25

Effective rate (APY): 12.55%

4 periods, $\dfrac{12\%}{4} = 3\%$

Complete the following using present value Table 12.3:

	Amount desired at end of period	Length of time	Rate	Compounded	On PVTable 12.3 Period used	On PVTable 12.3 Rate used	PV factor used	PV of amount desired at end of period
12–7.	$2,600	6 years	4%	Semiannually	12	2%	.7885	$2,050.10

$2,600 × .7885 = $2,050.10

`Excel`

WORD PROBLEMS

12–13. Jean Rich, owner of a local Dunkin' Donuts shop, loaned $14,000 to Mel Lyon to help him open an Internet business. Mel plans to repay Jean at the end of 6 years with 6% interest compounded semiannually. How much will Jean receive at the end of 6 years?

6 years × 2 = 12 periods $\dfrac{6\%}{2} = 3\%$ $14,000 × 1.4258 = $19,961.20

12–17. Lee Wills loaned Audrey Chin $16,000 to open a hair salon. After 6 years, Audrey will repay Lee with 8% interest compounded quarterly. How much will Lee receive at the end of 6 years?

6 years × 4 = 24 periods $\dfrac{8\%}{4} = 2\%$ $16,000 × 1.6084 = $25,734.40

12–21. St. Paul Federal Bank is quoting 1-year Certificates of Deposits with an interest rate of 5% compounded semiannually. Joe Saver purchased a $5,000 CD. What is the CD's effective rate (APY) to the nearest hundredth percent? Use tables in the *Business Math Handbook*.

1 year $\times$ 2 = 2 periods $\qquad \dfrac{5\%}{2} = 2\tfrac{1}{2}\%$ $\qquad$ Effective rate (APY) = $\dfrac{\$253}{\$5,000 \times 1} = 5.06\%$

$$\begin{array}{r} \$5,000 \\ \times\ 1.0506 \\ \hline \$5,253 \\ -\ \ 5,000 \\ \hline \$\ \ 253\ \text{interest} \end{array}$$

12–25. Pete Air wants to buy a used Jeep in 5 years. He estimates the Jeep will cost $15,000. Assume Pete invests $10,000 now at 12% interest compounded semiannually. Will Pete have enough money to buy his Jeep at the end of 5 years?

Compounding $\qquad\qquad\qquad\qquad$ **or** $\quad$ **Present value**

5 years $\times$ 2 = 10 periods $\qquad \dfrac{12\%}{2} = 6\%$ $\qquad$ 10 periods $\qquad$ $15,000 $\times$ .5584 = $8,376

$10,000 $\times$ 1.7908 = $17,908 $\qquad$ Yes. $\qquad$ 6% $\qquad\qquad\qquad\qquad\qquad\qquad\qquad$ Yes.

12–27. Paul Havlik promised his grandson Jamie that he would give him $6,000 8 years from today for graduating from high school. Assume money is worth 6% interest compounded semiannually. What is the present value of this $6,000?

8 years $\times$ 2 = 16 periods $\qquad \dfrac{6\%}{2} = 3\%$ $\qquad$ $6,000 $\times$ .6232 = $3,739.20

CHALLENGE PROBLEM

12–31. The U.S. government has ended 20 years of litigation by agreeing to pay $18 million to the estate of Richard M. Nixon. On June 14, 2000, the *Los Angeles Times* reported that the estate had demanded $35 million plus $8\tfrac{1}{2}\%$ interest compounded annually for items confiscated after Nixon resigned the presidency. **(a)** How much interest did the estate want? **(b)** What was the total amount the estate wanted? **(c)** How much did the government save by settling for $18 million?

a. 20 years $\times$ 1 = 20 periods

$\qquad 8\tfrac{1}{2}\% \div 1 = 8\tfrac{1}{2}\% = 5.1121 \times \$35,000,000$

$$\begin{array}{r} =\ \ \$178,923,500 \\ -\ \ \ 35,000,000 \\ \hline \$143,923,500\ \text{interest} \end{array}$$

b. $178,923,500 total amount

c.
$$\begin{array}{r} \$178,923,500 \\ -\ \ 18,000,000 \\ \hline \$160,923,500\ \text{government saved} \end{array}$$

END-OF-CHAPTER PROBLEMS

DRILL PROBLEMS

Complete the ordinary annuities for the following using tables in the *Business Math Handbook:*

	Amount of payment	Payment payable	Years	Interest rate	Value of annuity
13–1.	$5,000	Quarterly 20	5 22.0190	4% 1%	$110,095 ($5,000 × 22.0190)

Redo Problem 13–1 as an annuity due:

13–3. $5,000, 21 periods, 1% = $5,000 × 23.2392 = $116,196
$$\begin{array}{r} -\quad 5,000 \\ \hline \$111,196 \end{array}$$

Calculate the value of the following annuity due without a table. Check your results by Table 13.1 or the *Business Math Handbook* (they will be slightly off due to rounding):

	Amount of payment	Payment payable	Years	Interest rate
13–4.	$2,000	Annually	3	6%

$2,000.00	
+ 120.00	
$2,120.00	$4,367.20
+ 2,000.00	+ 2,000.00
$4,120.00	$6,367.20
+ 247.20	+ 382.03
$4,367.20	$6,749.23

Check 4 periods, 6%
$2,000 × 4.3746 = $8,749.20
$$\begin{array}{r} -\ 2,000.00 \\ \hline \$6,749.20 \end{array}$$

Complete the following, using Table 13.2 or the *Business Math Handbook* for the present value of an ordinary annuity:

	Amount of annuity expected	Payment	Time	Interest rate	Present value (amount needed now to invest to receive annuity)
13–5.	$900	Annually	4 years	6%	$3,118.59 ($900 × 3.4651) (4 periods, 6%)

13–7. Check Problem 13–5 without the use of Table 13.2.

	$3,118.59	$2,405.71	$1,650.05	$849.05
($3,118.59 × .06)	+ 187.12	+ 144.34	+ 99.00	+ 50.94*
	$3,305.71	$2,550.05	$1,749.05	$899.99
	− 900.00	− 900.00	− 900.00	− 900.00
	$2,405.71	$1,650.05	$ 849.05	$ .00

*Off 1 cent due to rounding.

Using the sinking fund Table 13.3 or the *Business Math Handbook,* complete the following:

	Required amount	Frequency of payment	Length of time	Interest rate	Payment amount end of each period
13–8.	$25,000	Quarterly	6 years	8%	$822.50 (24 periods, 2% = .0329) $25,000 × .0329 = $822.50
13–9.	$15,000	Annually	8 years	8%	$1,410 (8 periods, 8% = .0940) $15,000 × .0940 = $1,410

WORD PROBLEMS (Use Tables in the *Business Math Handbook*)

13–13. On April 12, 2001, Terry Savage of the *Chicago Sun-Times* wrote a column on saving for retirement. Assuming the stock market earns the historical market average return of 10.5%, you will have a nice retirement nest egg if you save $5,000 a year for 31 years. What is the value of this ordinary annuity?

31 periods, $10\frac{1}{2}$%

$$\begin{array}{r} 200.8741 \\ \times \$ \quad 5,000.00 \\ \hline \$1,004,370.50 \end{array}$$

13–15. The October 2000 issue of *Black Enterprise* reported on compounding. If you were able to invest only $1,200 at the end of each quarter and placed it in a vehicle that produced an average annual return of 6% compounded quarterly, how much would you receive in 10 years?

10 years × 4 = 40 periods

6% ÷ 4 = $1\frac{1}{2}$%

$$\begin{array}{r} 54.2677 \\ \times \$ \quad 1,200 \\ \hline \$65,121.24 \end{array}$$

13–21. At the beginning of each period for 10 years, Merl Agnes invests $500 semiannually at 6%. What is the cash value of this annuity due at the end of year 10?

20 periods + 1 = 21, 3% $500 × 28.6765 =

$$\begin{array}{r} \$14,338.25 \\ - \quad 500.00 \\ \hline \$13,838.25 \end{array}$$

13–23. On Joe's graduation from college, Joe Martin's uncle promised him a gift of $12,000 in cash or $900 every quarter for the next 4 years after graduation. If money could be invested at 8% compounded quarterly, which offer is better for Joe?

`Excel`

16 periods, $\dfrac{8\%}{4}$ = 2% $900 × 13.5777 = $12,219.93 **or** $900 × 18.6392 = $16,775.28

(Table 13.2)   (Table 13.1) $\begin{array}{r} \times \quad .7284 \text{ (Table 12.3)} \\ \hline \$12,219.11 \end{array}$

Choose the annuity.

2%
16 periods

13–25. GU Corporation must buy a new piece of equipment in 5 years that will cost $88,000. The company is setting up a sinking fund to finance the purchase. What will the quarterly deposit be if the fund earns 8% interest?

20 periods, 2% (Table 13.3)

.0412 × $88,000 = $3,625.60 quarterly payment

CHALLENGE PROBLEM

13–31. Ajax Corporation has hired Brad O'Brien as its new president. Terms included the company's agreeing to pay retirement benefits of $18,000 at the end of each semiannual period for 10 years. This will begin in 3,285 days. If the money can be invested at 8% compounded semiannually, what must the company deposit today to fulfill its obligation to Brad?

10 years $\times$ 2 = 20 periods $\qquad \dfrac{8\%}{2} = 4\%$

$18,000 $\times$ 13.5903 = $244,625.40 (Table 13.2)

$\dfrac{3,285 \text{ days}}{365 \text{ days per year}} = 9$ years $\qquad$ 9 years $\times$ 2 = 18 periods $\qquad \dfrac{8\%}{2} = 4\%$

$244,625.40 $\times$ .4936 = $120,747.09

Check $120,747.09 $\times$ 2.0258 = $244,609.45 (off due to rounding)

1. Amy O'Mally graduated from high school. Her uncle promised her as a gift a check for $2,000 or $275 every quarter for 2 years. If money could be invested at 6% compounded quarterly, which offer is better for Amy? (Use the tables in the *Business Math Handbook*.) *(p. 310)*

   2 years $\times$ 4 = 8 periods $\dfrac{6\%}{4} = 1\frac{1}{2}\%$

   $275 $\times$ 7.4859 = $2,058.62   **or** $275 $\times$ 8.4328 = $2,319.02
   Take the annuity. $\times$ .8877
                                                        $2,058.59

3. Roger Disney decides to retire to Florida in 12 years. What amount should Roger invest today so that he will be able to withdraw $30,000 at the end of each year for 20 years *after* he retires? Assume he can invest money at 8% interest compounded annually. (Use tables in the *Business Math Handbook*.) *(p. 306)*
   20 periods, 8%   (Table 13.2)   9.8181 $\times$ $30,000 = $294,543
   12 periods, 8%   (Table 12.3)   .3971 $\times$ $294,543 = $116,963.02

5. Sue Cooper borrowed $6,000 on an $11\frac{3}{4}\%$, 120-day note. Sue paid $300 toward the note on day 50. On day 90, Sue paid an additional $200. Using the U.S. Rule, Sue's adjusted balance after her first payment is the following. *(p. 249)*

   $6,000 $\times$ .1175 $\times$ $\dfrac{50}{360}$ = $97.92 $300.00 $6,000 − $202.08 = $5,797.92
                                    −   97.92
                                    $202.08

7. Alice Reed deposits $16,500 into Rye Bank, which pays 10% interest compounded semiannually. Using the appropriate table, what will Alice have in her account at the end of 6 years? *(p. 290)*

   6 years $\times$ 2 = 12 periods $\dfrac{10\%}{2} = 5\%$ (Table 12.1)

   12 periods, 5% = 1.7959 $16,500 $\times$ 1.7959 = $29,632.35

END-OF-CHAPTER PROBLEMS

DRILL PROBLEMS

Complete the following table:

	Purchase price of product	Down payment	Amount financed	Number of monthly payments	Amount of monthly payments	Total of monthly payments	Total finance charge
14–1. Chrysler PT Cruiser	$22,500 −	$6,000 =	$16,500	60 ×	$310 =	$18,600	$2,100 ($18,600 − $16,500)

Calculate **(a)** the amount financed, **(b)** the total finance charge, and **(c)** APR by table lookup.

	Purchase price of a used car	Down payment	Number of monthly payments	Amount financed	Total of monthly payments	Total finance charge	APR
14–3.	$5,673	$1,223	48	$4,450	$5,729.76	$1,279.76	12.75%–13%

$$\begin{array}{r} \$5,673 \\ -\ 1,223 \\ \hline \$4,450 \end{array} \qquad \begin{array}{r} \$5,729.76 \\ -\ 4,450.00 \\ \hline \$1,279.76 \end{array} \qquad \frac{\$1,279.76}{\$4,450.00} \times \$100 = \$28.76 \text{ is between 12.75\% and 13\% at 48 months}$$

Calculate the monthly payment for Problems 14–3 and 14–4 by table lookup and formula. (Answers will not be exact due to rounding of percents in table lookup.)

14–5. **(14–3)** (Use 13% for table lookup.)

Table:
$$\begin{array}{r} \$5,673 \\ -\ 1,223 \\ \hline \$4,450 \end{array} \div \$1,000 = \begin{array}{r} 4.45 \\ \times\ 26.83 \ (13\%, 48 \text{ months}) \\ \hline \$119.39 \end{array}$$

Formula: $\dfrac{\$1,279.76 + \$4,450}{48} = \$119.37$

Calculate the finance charge rebate and payoff:

	Loan	Months of loan	End-of-month loan is repaid	Monthly payment	Finance charge rebate	Final payoff
14–7.	$7,000	36	10	$210	$295.14	$5,164.86

Step 1.
Total payments
$$\begin{array}{rr} 36 \times \$210 = & \$7,560 \\ 10 \times \$210 = & -\ 2,100 \\ \hline \text{Balance outstanding} & \$5,460 \end{array}$$

Step 2.
$$\begin{array}{lr} \text{Total of all payments } 36 \times \$210 & \$7,560 \\ \text{Amount financed} & -\ 7,000 \\ \hline \text{Total finance charge} & \$\ 560 \end{array}$$

Step 3. $36 − 10 = 26$

Step 4. By Table 14.3 $\dfrac{351}{666} \begin{array}{l} \longrightarrow 26 \text{ months to go} \\ \longrightarrow 36 \text{ months total loan} \end{array}$

Step 5. $\dfrac{351}{666} \times \$560 = \$295.14$ finance charge rebate

(Step 4) (Step 2)

Step 6. $\$5,460 − \$295.14 = \$5,164.86$ (payoff)
(Step 1) (Step 5)

14–9. Calculate the average daily balance:

30-day billing cycle			
9/16	Billing date	Previous balance	$2,000
9/19	Payment	$ 60	
9/30	Charge: Home Depot	1,500	
10/3	Payment	60	
10/7	Cash advance	70	

No. of days of current balance	Current balance	Extension
3	$2,000	$ 6,000
11	1,940	21,340
3	3,440	10,320
4	3,380	13,520
9	3,450	31,050

$$\frac{\$82,230}{30} = \$2,741 \text{ average daily balance}$$

WORD PROBLEMS

14–11. The *Edmunds 2001 Buyer's Guide* listed a 2001 Ford Taurus SE wagon at $18,646. Margaret Paine purchased the car with $1,900 down. The loan is for 48 months with monthly payments of $408.10. **(a)** What amount did Margaret finance? **(b)** What is her finance charge? **(c)** Calculate the deferred payment price. **(d)** What is Margaret's APR?

a. $18,646 − $1,900 = $16,746.00 amount financed

b. $19,588.80 ($408.10 × 48) − $16,746.00 = $2,842.80 finance charge

c. $19,588.80 ($408.10 × 48) + $1,900 = $21,488.80 deferred payment price

d. $\dfrac{\$2,842.80}{\$16,746.00} \times \$100 = 16.97599 = 16.98$ (Table 14.1, between 7.75% and 8.00%)

14–13. From this partial advertisement calculate:

$95.10 per month
#43892 Used car. Cash price $4,100. Down payment $50. For 60 months.

a. Amount financed. b. Finance charge.
c. Deferred payment price. d. APR by Table 14.1.
e. Check monthly payment (by formula).

a. Amount financed = $4,100 − $50 = $4,050
b. Finance charge = $5,706 ($95.10 × 60) − $4,050 = $1,656
c. Deferred payment price = $5,706 ($95.10 × 60) + $50 = $5,756
d. $\dfrac{\$1,656}{\$4,050} \times \$100 = \$40.89$ (Table 14.1, between 14.25% and 14.50%)
e. $\dfrac{\$1,656 + \$4,050}{60} = \$95.10$

14–17. On June 24, 2001, the *Bank Rate Monitor* reported on auto loan rates. First America Bank's monthly payment charge on a 48-month $20,000 loan is $488.26. The U.S. Bank's monthly payment fee is $497.70 for the same loan amount. What would be the APR for each of these banks?

First America Bank

$488.26 × 48 = $23,436.48
              − 20,000.00
              $ 3,436.48 finance charge

$\dfrac{\$3,436.48}{\$20,000} \times \$100 = 17.1824$

= Between 8.00% and 8.25%

U.S. Bank

$497.70 × 48 = $23,889.60
              − 20,000.00
              $ 3,889.60 finance charge

$\dfrac{\$3,889.60}{\$20,000} \times \$100 = 19.448$

= Between 8.75% and 9%

CHALLENGE PROBLEM

14–21. You have a $1,100 balance on your 15% credit card. You have lost your job and been unemployed for 6 months. You have been unable to make any payments on your balance. However, you received a tax refund and want to pay off the credit card. How much will you owe on the credit card, and how much interest will have accrued? What will be the effective rate of interest after the 6 months (to nearest hundredth percent)?

Interest

1 month: $1,100.00 $\times$ .15 $\times \frac{1}{12}$ = $13.75 = $1,113.75

2 months: $1,113.75 $\times$ .15 $\times \frac{1}{12}$ = $13.92 = $1,127.67

3 months: $1,127.67 $\times$ .15 $\times \frac{1}{12}$ = $14.10 = $1,141.77

4 months: $1,141.77 $\times$ .15 $\times \frac{1}{12}$ = $14.27 = $1,156.04

5 months: $1,156.04 $\times$ .15 $\times \frac{1}{12}$ = $14.45 = $1,170.49

6 months: $1,170.49 $\times$ .15 $\times \frac{1}{12}$ = $14.63 = $1,185.12

$$\frac{\$85.12 \text{ interest}}{\$1,100 \times \frac{6}{12}} = \frac{\$85.12}{\$550} = 15.48\%$$

END-OF-CHAPTER PROBLEMS

DRILL PROBLEMS

Complete the following amortization chart by using Table 15.1.

	Selling price of home payment	Down payment	Principal (loan)	Rate of interest	Years	Payment per $1,000	Monthly mortgage
15–1. Excel	$159,000	$10,000	$149,000	$6\frac{1}{2}$%	25	$6.76	$1,007.24 (149 × $6.76)

Complete the following:

	Selling price	Down payment	Amount mortgage	Rate	Years	Monthly payment	First payment broken down into— Interest	First payment broken down into— Principal	Balance at end of month
15–7.	$199,000	$40,000	$159,000	$7\frac{1}{2}$%	12	$1,679.04	$1,656.25	$22.79	$158,977.21

$159 \times \$10.56 = \$1,679.04; \$159,000 \times .125 \times \frac{1}{12} = \$1,656.25$ ($159,000 − $22.79)

WORD PROBLEMS

15–9. On March 23, 2000, the *Philadelphia Inquirer* reported on 15-year versus 30-year fixed rate mortgages. The total interest for 15 years would be $72,000 and $164,000 for 30 years; this is based on a $100,000 loan at 8%. Doug Tweeten wants to take advantage of this savings. **(a)** How much would his monthly payments be on a 15-year loan? **(b)** How much would his monthly payments be on a 30-year loan? **(c)** How much more would he pay each month on a 15-year loan?

a. $\frac{\$100,000}{\$1,000} = 100 \times \overset{(15 \text{ years } 8\%)}{\$9.56} = \$956.00$

c. $956.00 monthly payment (15 years)
− 734.00 monthly payment (30 years)
$222.00 more for a 15-year loan each month

b. $\frac{\$100,000}{\$1,000} = 100 \times \overset{(30 \text{ years } 8\%)}{\$7.34} = \$734.00$

15–11. Bill Allen bought a home in Arlington, Texas, for $108,000. He put down 25% and obtained a mortgage for 30 years at 11%. What is Bill's monthly payment? What is the total interest cost of the loan?

$\$108,000 − \$27,000 = \frac{\$81,000}{\$1,000} = 81 \times \$9.53 = \$771.93$

$\$771.93 \times 360 = \$277,894.80 − \$81,000.00 = \$196,894.80$ total interest

15–13. Mike Jones bought a new split-level home for $150,000 with 20% down. He decided to use Victory Bank for his mortgage. They were offering $13\frac{3}{4}$% for 25-year mortgages. Provide Mike with an amortization schedule for the first three periods.

Payment number	Portion to— Interest	Portion to— Principal	Balance of loan outstanding	Monthly payment is:
1	$1,375.00 ($1,422 − $1,375)	$47.00 ($120,000 − $47.00)	$119,953.00	$\frac{\$120,000}{\$1,000} = 120 \times \$11.85 = \$1,422$ $\$120,000 \times .1375 \times \frac{1}{12} = \$1,375.00$
2	$1,374.46 ($1,422 − $1,374.46)	$47.54 ($119,953 − $47.54)	$119,905.46	$\$119,953.00 \times .1375 \times \frac{1}{12} = \$1,374.46$
3	$1,373.92	$48.08	$119,857.38	$\$119,905.46 \times .1375 \times \frac{1}{12} = \$1,373.92$

CHALLENGE PROBLEM

15–17. On December 21, 2000, the *Los Angeles Times* provided a chart to enable readers to figure out how refinancing would effect their mortgage payments. To use this chart, drop the last three zeros from your current mortgage amount and multiply that by the multiplier that most closely corresponds to your new interest rate and loan term. Subtract that number from your current loan payment, and that's your monthly savings.

Multipliers

Loan rate	30-year loan	15-year loan
8.0%	7.3376	9.5565
7.9%	7.2680	9.4988
7.8%	7.1987	9.4414
7.7%	7.1296	9.3841
7.6%	7.0607	9.3270
7.5%	6.9921	9.2701
7.4%	6.9238	9.2134

Consider a $200,000 loan at 8% with a monthly payment of $1,467.53. Calculate refinancing at 7.4% for 30 years. **(a)** What would be the savings per month? **(b)** If the loan costs were $2,000, how many months would it take for refinancing to pay off?

7.4% for 30 years $200 \times 6.9238 = \$1,384.76$ new monthly payment

a.
$1,467.53
− 1,384.76
$ 82.77 saved each month

b. $\dfrac{\$2,000}{\$82.77} = 24.163344$ Almost 25 months.

END-OF-CHAPTER PROBLEMS

DRILL PROBLEMS

16–1. As the accountant for True Value Hardware, prepare a December 31, 2004, balance sheet like that for The Card Shop (LU 16–1) from the following: cash, $15,000; accounts payable, $18,000; merchandise inventory, $7,000; A. Long, capital, $18,000; and equipment, $14,000.

TRUE VALUE HARDWARE			
Balance Sheet			
December 31, 2004			
Assets		**Liabilities**	
Cash	$15,000	Accounts payable	$18,000
Merchandise inventory	7,000	**Owner's Equity**	
Equipment	14,000	A. Long, capital	18,000
Total assets	$36,000	Total liabilities and owner's equity	$36,000

16–7. Complete the comparative income statement and balance sheet for Logic Company (round percents to the nearest hundredth):

LOGIC COMPANY Comparative Income Statement For Years Ended December 31, 2004 and 2005				
			Increase (decrease)	
	2005	**2004**	**Amount**	**Percent**
Gross sales	$19,000	$15,000	$4,000	26.67
Sales returns and allowances	1,000	100	900	900.00
Net sales	$18,000	$14,900	$3,100	+ 20.81
Cost of merchandise (goods) sold	12,000	9,000	3,000	+ 33.33
Gross profit	$ 6,000	$ 5,900	$ 100	+ 1.69
Operating expenses:				
Depreciation	$ 700	$ 600	$ 100	+ 16.67
Selling and administrative	2,200	2,000	200	+ 10.00
Research	550	500	50	+ 10.00
Miscellaneous	360	300	60	+ 20.00
Total operating expenses	$ 3,810	$ 3,400	$ 410	+ 12.06
Income before interest and taxes	$ 2,190	$ 2,500	$ (310)	− 12.40
Interest expense	560	500	60	+ 12.00
Income before taxes	$ 1,630	$ 2,000	$ (370)	− 18.50
Provision for taxes	640	800	(160)	− 20.00
Net income	$ 990	$ 1,200	$ (210)	− 17.50

$\dfrac{\$3,100}{\$14,900}$

LOGIC COMPANY
Comparative Balance Sheet
December 31, 2004 and 2005

	2005 Amount	2005 Percent	2004 Amount	2004 Percent
Assets				
Current assets:				
Cash	$12,000	13.48	$ 9,000	13.74
Accounts receivable	16,500	18.54	12,500	19.08
Merchandise inventory	8,500	9.55	14,000	21.37
Prepaid expenses	24,000	26.97	10,000	15.27
Total current assets	$61,000	68.54	$45,500	69.47*
Plant and equipment:				
Building (net)	$14,500	16.29	$11,000	16.79
Land	13,500	15.17	9,000	13.74
Total plant and equipment	$28,000	31.46	$20,000	30.53
Total assets	$89,000	100.00	$65,500	100.00
Liabilities				
Current liabilities:				
Accounts payable	$13,000	14.61	$ 7,000	10.69
Salaries payable	7,000	7.87	5,000	7.63
Total current liabilities	$20,000	22.47*	$12,000	18.32
Long-term liabilities:				
Mortgage note payable	22,000	24.72	20,500	31.30
Total liabilities	$42,000	47.19	$32,500	49.62
Stockholders' Equity				
Common stock	$21,000	23.60	$21,000	32.06
Retained earnings	26,000	29.21	12,000	18.32
Total stockholders' equity	$47,000	52.81	$33,000	50.38
Total liabilities and stockholders' equity	$89,000	100.00	$65,500	100.00

Annotation next to Cash 2004: $\dfrac{\$9,000}{\$65,500}$

*Due to rounding.

From Problem 16–7, your supervisor has requested that you calculate the following ratios (round to the nearest hundredth):

	2005	2004
16–9. Acid test.	1.43	1.79
16–13. Net income (after tax) to the net sales.	.06	.08

16–9. $\dfrac{\text{CA} - \text{Inv} - \text{Prepaid expenses}}{\text{CL}}$ $\dfrac{\$61,000 - \$8,500 - \$24,000}{\$20,000} = 1.43$ $\dfrac{\$45,500 - \$14,000 - \$10,000}{\$12,000} = 1.79$

16–13. $\dfrac{\text{NI}}{\text{Net sales}}$ $\dfrac{\$990}{\$18,000} = .055 = .06$ $\dfrac{\$1,200}{\$14,900} = .0805 = .08$

WORD PROBLEMS

16–15. An Internet magazine contained an article that stated the following: "The net income for 2004 was $800,000, and the return on equity was 20." What was the amount of equity to the nearest dollar?

$$\text{Return on equity} = \frac{\text{Net income}}{\text{Equity}}$$

$$\frac{\$800,000 \text{ (net income)}}{?} = 20\%$$

$$\frac{\$800,000}{.20} = \$4,000,000$$

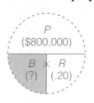

16–17. Find the following ratios for Motorola Credit Corporation from the Motorola 2000 annual report: **(a)** total debt to total assets, **(b)** return on equity, **(c)** asset turnover, **(d)** profit margin on net sales. Round to the nearest hundredth.

	2000 (dollars in millions)
Net revenue (sales)	$ 265
Net earnings	147
Total assets	2,015
Total liabilities	1,768
Total stockholders' equity	427

a. $\dfrac{\text{Total liabilities}}{\text{Total assets}}$ $\dfrac{\$1,768}{\$2,015} = 87.74\%$

b. $\dfrac{\text{Net income}}{\text{Stockholders' equity}}$ $\dfrac{\$147}{\$427} = 34.43\%$

c. $\dfrac{\text{Net sales}}{\text{Total assets}}$ $\dfrac{\$265}{\$2,015} = 13¢$

d. $\dfrac{\text{Net income}}{\text{Net sales}}$ $\dfrac{\$147}{\$265} = 55.47\%$

CHALLENGE PROBLEM

16–20. On March 20, 2001, the *Houston Chronicle* reported on the profits of Oshman's Sporting Goods. For the fourth quarter, Oshman's reported net income of $7.3 million. That compares with a net income of $1.5 million the same quarter a year earlier. Net sales for the quarter increased 12.5% to $105.4 million. **(a)** What was Oshman's profit margin on net sales this quarter? **(b)** What was the percent increase in net income? **(c)** What were last year's net sales? **(d)** What was last year's profit margin on net sales? Round to the nearest hundredth percent.

a. $\dfrac{\text{Net income}}{\text{Net sales}} = \dfrac{\$7.3 \text{ million}}{\$105.4 \text{ million}} = 6.9259962\%$

$= 6.93\%$

b.
$$\begin{array}{r} \$7.3 \text{ million} \\ - \ 1.5 \text{ million} \\ \hline \$5.8 \text{ million increase} \end{array}$$
$\dfrac{\$5.8 \text{ million}}{\$1.5 \text{ million}} = 386.66666\%$

$= 386.67\%$

c. $\dfrac{\$105.4 \text{ million}}{1.125 \ (100\% + 12.5\%)} = \$93.68888 = \$93.69 \text{ million}$

d. $\dfrac{\$1.5 \text{ million}}{\$93.69 \text{ million}} = 1.6010246\% = 1.60\%$

END-OF-CHAPTER PROBLEMS

DRILL PROBLEMS

From the following facts, complete a depreciation schedule, using the straight-line method:

Given   Cost of Range Rover $50,000
Residual value   10,000
Estimated life   8 years

End of year	Cost of Range Rover	Depreciation expense for year	Accumulated depreciation at end of year	Book value at end of year
17–1. 1	$50,000	$5,000	$5,000	$45,000
17–7. 7	$50,000	$5,000	$35,000	$15,000

$$($50,000 - $5,000)\ \frac{$50,000 - $10,000}{8 \text{ years}} = $5,000$$

Prepare a depreciation schedule using the sum-of-the-years'-digits method:

Given   Ford pickup $11,000
Residual value   1,000
Estimated life   4 years

End of year	Cost − Residual value	×	Fraction for year	=	Depreciation expense for year	Accumulated depreciation at end of year	Book value at end of year
17–9. 1	$10,000	×	$\frac{4}{10}$		$4,000	$ 4,000	$7,000 ($11,000 − $4,000)
17–10. 2	$10,000	×	$\frac{3}{10}$		$3,000	$ 7,000	$4,000
17–11. 3	$10,000	×	$\frac{2}{10}$		$2,000	$ 9,000	$2,000
17–12. 4	$10,000	×	$\frac{1}{10}$		$1,000	$10,000	$1,000

$$\frac{N(N + 1)}{2} = \frac{4(4 + 1)}{2} = \frac{20}{2} = 10$$

WORD PROBLEMS

17–25. Pat Brown bought a Chevy truck for $28,000 with an estimated life of 5 years. The residual value of the truck is $3,000. Assume a straight-line method of depreciation. **(a)** What will be the book value of the truck at the end of year 3? **(b)** If the Chevy truck was bought the first year on April 12, how much depreciation would be taken the first year?

a. $\frac{$28,000 - $3,000}{5 \text{ years}} = $5,000$ depreciation expense per year

   $28,000 − $15,000 = $13,000 book value

b. $$5,000 \times \frac{9}{12} = $3,750$

17–29. Ray Kunz, owner of Ray's Auto Service, purchased a 2001 Corvette convertible listed in *Edmunds 2001 Buyer's Guide* at $46,605 with a residual value of $24,235. Ray uses the Corvette in his business. The life expectancy is 5 years. Using the declining-balance method, what would the depreciation expense be the first year?

$46,605 × .40 (twice straight line) = $18,642 depreciation first year

CHALLENGE PROBLEM

17–33. A piece of equipment was purchased July 26, 2003, at a cost of $72,000. The estimated residual value is $5,400 with a useful life of 5 years. Assume a production life of 60,000 units. Compute the depreciation for years 2003 and 2004 using **(a)** straight-line; **(b)** units-of-production (in 2003, 5,000 units produced and in 2004, 18,000 units produced); and **(c)** sum-of-the-years'-digits methods.

a. $\dfrac{\$72,000 - \$5,400}{5 \text{ years}} = \$13,320$

2003: $\$13,320 \times \dfrac{5}{12} = \$5,550^{(5/12 \text{ Aug. to Dec. 31})}$

2004: $\$13,320$

b. $\dfrac{\$72,000 - \$5,400}{60,000} = \$1.11$

2003: $5,000 \times \$1.11 = \$5,550$

2004: $18,000 \times \$1.11 = \$19,980$

c. 2003: $\$66,600^{(\$72,000 - \$5,400)} \times \dfrac{5}{15} = \$22,200 \longrightarrow$ 2003: $\dfrac{5}{12} \times \$22,200 = \$9,250$

2004: $\$66,600 \times \dfrac{4}{15} = \$17,760 \longrightarrow$ 2004: $\dfrac{7}{12} \times \$22,200 = 12,950$

$\dfrac{5}{12} \times \$17,760 = \dfrac{7,400}{\$20,350}$

END-OF-CHAPTER PROBLEMS

DRILL PROBLEMS

18–1. Using the specific identification method, calculate **(a)** the cost of ending inventory and **(b)** the cost of goods sold given the following:

Date	Units purchased	Cost per scooter		Ending inventory
July 1	50 Razor scooters	$20	$1,000	6 scooters from July 1
September 1	70 Razor scooters	30	2,100	9 scooters from September 1
November 1	90 Razor scooters	35	3,150	12 scooters from November 1
			$6,250	

July 1	6 scooters × $20 = $120	$6,250 cost of goods available for sale
September 1	9 scooters × $30 = 270	− 810 **(a)** cost of ending inventory
November 1	12 scooters × $35 = 420	$5,440 **(b)** cost of goods sold
	$810	

From the following, **(a)** calculate the cost of ending inventory (round the average unit cost to the nearest cent) and **(b)** cost of goods sold using the weighted-average method, FIFO, and LIFO (ending inventory shows 61 units).

	Number purchased	Cost per unit	Total
January 1 inventory	40	$4	$ 160
April 1	60	7	420
June 1	50	8	400
November 1	55	9	495
	205		$1,475

18–3. FIFO:

$$55 × \$9 = \$495$$
$$6 × 8 = + 48$$

a. Cost of ending inventory $543 FIFO—old sold first.

   Cost of goods available for sale $1,475

   Ending inventory − 543

b. Cost of goods sold $ 932

18–17. Complete the following (round answers to the nearest hundredth):

a. Average inventory at cost	b. Average inventory at retail	c. Net sales	d. Cost of goods sold	e. Inventory turnover at cost	f. Inventory turnover at retail
$14,000	$21,540	$70,000	$49,800	3.56	3.25

e. $= \dfrac{\$49,800 \text{ (d)}}{\$14,000 \text{ (a)}} = 3.56$ f. $= \dfrac{\$70,000 \text{ (c)}}{\$21,540 \text{ (b)}} = 3.25$

Complete the following (assume $90,000 of overhead to be distributed):

		Square feet	Ratio	Amount of overhead allocated
18–18.	Department A	10,000	.25 (10,000 ÷ 40,000)	$22,500 (.25 × $90,000)
18–19.	Department B	30,000	.75 (30,000 ÷ 40,000)	$67,500 (.75 × $90,000)

WORD PROBLEMS

18–23. The May 2001 issue of *Tax Adviser* reported on tax consequences using LIFO versus FIFO methods. Mountain State Ford used the LIFO inventory method for its parts and accessories (parts) inventory. The following purchases were made: 60 spark plugs at $.70 each, 40 spark plugs at $.73 each, 45 spark plugs at $.75 each, and 30 spark plugs at $.80 each. A recent inventory indicated 85 spark plugs were still in stock. What would be the cost of the ending inventory using LIFO and FIFO?

			LIFO			**FIFO**		
60 at $.70 =	$ 42.00		60 at $.70 =	$42.00		30 at $.80 =	$24.00
40 at $.73 =	29.20		25 at $.73 =	+ 18.25		45 at .75 =	33.75
45 at $.75 =	33.75		Ending inventory	$60.25		10 at .73 =	7.30
30 at $.80 =	24.00					Ending inventory	$65.05
175	$128.95						

18–25. May's Dress Shop's inventory at cost on January 1 was $39,000. Its retail value is $59,000. During the year, May purchased additional merchandise at a cost of $195,000 with a retail value of $395,000. The net sales at retail for the year were $348,000. Could you calculate May's inventory at cost by the retail method? Round the cost ratio to the nearest whole percent.

	Cost	Retail
Beginning inventory	$ 39,000	$ 59,000
Purchases	195,000	395,000
Cost of goods available for sale	$234,000	$454,000
Less net sales for year		348,000
Ending inventory at retail		$106,000
Cost ratio ($234,000 ÷ $454,000)		52%
Ending inventory at cost (.52 × $106,000)	$ 55,120	

CHALLENGE PROBLEM

18–29. Logan Company uses a perpetual inventory system on a FIFO basis. Assuming inventory on January 1 was 800 units at $8 each, what is the cost of ending inventory at the end of October 5?

Received			**Sold**	
Date	**Quantity**	**Cost per unit**	**Date**	**Quantity**
Apr. 15	220	$5	Mar. 8	500
Nov. 12	1,900	9	Oct. 5	200

Jan. 1 inventory	$6,400	800 units at $8	Oct. 5 inventory	$1,900	100 units at $8
Mar. 8 inventory	2,400	300 units at $8			220 units at $5
Apr. 15 inventory	3,500	300 units at $8			
		220 units at $5			

END-OF-CHAPTER PROBLEMS

DRILL PROBLEMS

Calculate the following:

	Retail selling price	Sales tax (5%)	Excise tax (9%)	Total price including taxes
19–1.	$ 900 +	$45 ($900 × .05)	+ $81 ($900 × .09)	$1,026

Calculate the actual sales since the sales and sales tax were rung up together; assume a 6% sales tax (round your answer to the nearest cent):

19–3. $\dfrac{\$90,000}{1.06} = \$84,905.66$

Complete the following:

	Tax rate per dollar	In percent	Per $100	Per $1,000	Mills
19–9.	.0699	6.99% (.0699)	$6.99 (.0699 × 100)	69.90 (.0699 × 1,000)	69.90 $\left(\dfrac{.0699}{.001}\right)$

WORD PROBLEMS

19–15. Don Chather bought a new computer for $1,995. This included a 6% sales tax. What is the amount of sales tax and the selling price before the tax?

$\dfrac{\$1,995}{1.06} = \$1,882.08$ actual sale $1,995 - \$1,882.08 = \$112.92$ sales tax

19–21. Bill Shass pays a property tax of $3,200. In his community, the tax rate is 50 mills. What is Bill's assessed value?

Mills × .001 × A = $ 3,200
50 × .001 × A = $ 3,200
.05A = $ 3,200
A = $64,000

19–23. On May 6, 2000, the *Daily Herald* reported on a school district losing funds on two local businesses because of proposed reduced property tax assessments. Bake Line Products is seeking to have its property reassessed from the $2.27 million, at which it's now valued, to $1.98 million. The owners of the Golf Center hope to have their property reassessed from $848,730 to $593,575. The current tax rate is .0725. How much will the district lose if both requests are granted? Round to the nearest ten thousands.

Bake Line	Golf	
$2,270,000	$848,730	$ 290,000
− 1,980,000	− 593,575	+ 255,155
$ 290,000 reduction	$255,155 reduction	$ 545,155 total reduction
		× .0725 tax rate
		$39,523.737 = $40,000 loss of taxes

CHALLENGE PROBLEM

19–27. Art Neuner, an investor in real estate, bought an office condominium. The market value of the condo was $250,000 with a 70% assessment rate. Art feels that his return should be 12% per month on his investment after all expenses. The tax rate is $31.50 per $1,000. Art estimates it will cost $275 per month to cover general repairs, insurance, and so on. He pays a $140 condo fee per month. All utilities and heat are the responsibility of the tenant. Calculate the monthly rent for Art. Round your answer to the nearest dollar (at intermediate stages).

$250,000 \times .70 = \$175,000$ assessed value

$\text{Tax} = 175 \times \$31.50 = \quad \$\ 5,512.50$ tax

$\qquad\qquad\qquad +\quad 3,300.00\ (\$275 \times 12)$ repairs and insurance

$\qquad\qquad\qquad \underline{+\quad 1,680.00}\ (\$140 \times 12)$ condo fee

$\qquad\qquad\qquad \quad \$10,492.50 \div 12 = \$874 \qquad\qquad \$874 \times 1.12 = \$978.88 = \$979$

END-OF-CHAPTER PROBLEMS

DRILL PROBLEMS

Calculate the annual premium for the following policies using Table 20.1 (for females subtract 3 years from the table).

Amount of coverage (face value of policy)	Age and sex of insured	Type of insurance policy	Annual premium
20–1. $70,000	26 F	Straight life	70 × $6.60 = $462

Calculate the following nonforfeiture options for Lee Chin, age 42, who purchased a $200,000 straight life policy. At the end of year 20, Lee stopped paying premiums.

20–7. Option 3: Extended term insurance
21 years 300 days

Calculate the short-rate premium and refund of the following:

Annual premium	Canceled after	Short-rate premium	Refund
20–9. $700	8 months by insured	$518 (.74 × $700)	$182 ($700 − $518)

Calculate the annual auto insurance premium for the following:

20–13. Britney Sper, Territory 5
Class 17 operator
Compulsory, 10/20/5 $ 258 ($98 + $160)

   Optional
   a. Bodily injury, 500/1,000 $ 298

   b. Property damage, 25M $ 166

   c. Collision, $100 deductible $ 233 ($190 + $43)
      Age of car is 2; symbol of car is 7

   d. Comprehensive, $200 deductible $ 112 ($108 + $4)

      Total annual premium $1,067

WORD PROBLEMS

20–19. Abby Ellen's toy store is worth $400,000 and is insured for $200,000. Assume an 80% coinsurance clause and that a fire caused $190,000 damage. What is the liability of the insurance company?

$$\frac{\$200,000}{\$320,000} \times \$190,000 = \$118,750 \qquad (.80 \times \$400,000)$$

20–21. The March 26, 2001, issue of *National Underwriter/Life & Health Financial Services* reported that auto insurance quotes gathered online could vary by as much as 300% on the same risk. Variations from $947 to $1,558 were given via the Internet. A class 18 operator carries compulsory 10/20/5 insurance. He has the following optional coverage: bodily injury, 500/1,000; property damage, 50M; and collision, $200 deductible. His car is 1 year old, and the symbol of the car is 8. He has comprehensive insurance, with a $200 deductible. Using your text, what is the total annual premium?

Class 18 operator

Compulsory, 10/20/5	$ 240	($80 + $160)
Optional		
Bodily injury, 500/1,000	251	
Property damage, 50M	168	
Collision, $200 deductible	280	($264 + $16)
Comprehensive, $200 deductible	161	($157 + $4)
Total annual premium	$1,100	

20–27. Marika Katz bought a new Blazer and insured it with only compulsory insurance 10/20/5. Driving up to her summer home one evening, Marika hit a parked car and injured the couple inside. Marika's car had damage of $7,500, and the car she struck had damage of $5,800. After a lengthy court suit, the couple struck were awarded personal injury judgments of $18,000 and $9,000, respectively. What will the insurance company pay for this accident, and what is Marika's responsibility?

Insurance company pays		**Marika pays**	
Bodily	$10,000 + $9,000	$ 8,000	
Property	5,000	7,500	no collision
		800	property damage not covered by compulsory
Total	$24,000	$16,300	

 CHALLENGE PROBLEM

20–29. Bill, who understands the types of insurance that are available, is planning his life insurance needs. At this stage of his life (age 35), he has budgeted $200 a year for life insurance premiums. Could you calculate for Bill the amount of coverage that is available under straight life and for a 5-year term? Could you also show Bill that if he were to die at age 40, how much more his beneficiary would receive if he'd been covered under the 5-year term? Round to the nearest thousand.

Straight life
$200 ÷ $11.26 = 17.762 × $1,000 = $17,762 = $18,000

$90,000
− 18,000
$72,000

Five-year term
$200 ÷ $2.23 = 89.686 × $1,000 = $89,686 = $90,000

END-OF-CHAPTER PROBLEMS

DRILL PROBLEMS

Calculate the cost (omit commission) of buying the following shares of stock:

21–1. 600 shares of America Online at $37.66. $22,596   (600 × $37.66)

Calculate the yield of each of the following stocks (round to the nearest tenth percent):

Company	Yearly dividend	Closing price per share	Yield
21–3. Boeing	$.68	$64.63	$1.1\% \left(\dfrac{\$.68}{\$64.63} \right)$

Calculate the earnings per share, price-earnings ratio (to nearest whole number), or stock price as needed:

Company	Earnings per share	Closing price per share	Price-earnings ratio
21–5. BellSouth	$3.15	$ 40.13	$13 \qquad \left(\dfrac{\$40.13}{\$3.15} \right)$

21–7. Calculate the total cost of buying 400 shares of CVS at $59.38. Assume a 2% commission.
400 shares × $59.38 = $23,752 × 1.02 = $24,227.04

21–9. Given: 20,000 shares cumulative preferred stock ($2.25 dividend per share): 40,000 shares common stock. Dividends paid: 2003, $8,000; 2004, 0; and 2005, $160,000. How much will preferred and common receive each year?

Year	2003	2004	2005
Dividend paid	$ 8,000	–0–	$160,000
Preferred	$ 8,000	–0–	$37,000 + $45,000
	($37,000)	($45,000)	+ $45,000 = $127,000
Common	–0–	–0–	$160,000 – $127,000
			= $33,000

For the following bonds, calculate the total annual interest, total cost, and current yield (to the nearest tenth percent):

Bond	Number of bonds purchased	Selling price	Total annual interest	Total cost	Current yield
21–13. Wang $6\frac{1}{2}$ 07	4	$68\frac{1}{8}$	$260.00	$2,725	9.5%

$.065 \times \$1,000 = \$65$
interest per bond

$\begin{array}{r} \$65 \\ \hline \$681.25 \end{array}$ ← $(68\frac{1}{8}\% = 68.125\% = .68125 \times \$1,000)$

WORD PROBLEMS

21–19. Maytag Company earns $4.80 per share. Today the stock is trading at $59.25. The company pays an annual dividend of $1.40. Calculate **(a)** the price-earnings ratio (round to the nearest whole number) and **(b)** the yield on the stock (to the nearest tenth percent).

a. $PE = \dfrac{\$59.25}{\$4.80} = 12$
b. $Yield = \dfrac{\$1.40}{\$59.25} = 2.4\%$

21–21. The following bond was quoted in *The Wall Street Journal*:

Bonds	Curr. yld.	Vol.	Close	Net chg.
NY Tel $7\frac{1}{4}$ 11	7.2	10	$100\frac{7}{8}$	$+1\frac{1}{8}$

Five bonds were purchased yesterday, and 5 bonds were purchased today. How much more did the 5 bonds cost today (in dollars)?

Today: $5 \times \$1,008.75\ (1.00875 \times \$1,000) = \$5,043.75$

Yesterday: $\begin{array}{r} 100\frac{7}{8} \\ -\ 1\frac{1}{8} \\ \hline 99\frac{6}{8} = 99\frac{3}{4} \end{array}$

$5 \times \$997.50\ (.9975 \times 1,000) = \begin{array}{r} -\ 4,987.50 \\ \hline \$\quad 56.25 \end{array}$

21–27. Louis Hall read in the paper that Fidelity Growth Fund has a NAV of $16.02. He called Fidelity and asked them how the NAV was calculated. Fidelity gave him the following information:

Current market value of fund investment	$8,550,000
Current liabilities	$ 860,000
Number of shares outstanding	480,000

Did Fidelity provide Louis with the correct information?

Yes. $\dfrac{\$8,550,000 - \$860,000}{480,000} = \$16.02$

CHALLENGE PROBLEM

21–31. On September 6, Irene Westing purchased one bond of Mick Corporation at $98\frac{1}{2}$. The bond pays $8\frac{3}{4}$ interest on June 1 and December 1. The stockbroker told Irene that she would have to pay the accrued interest and the market price of the bond and a $6 brokerage fee. What was the total purchase price for Irene? Assume a 360-day year (each month is 30 days) in calculating the accrued interest. (*Hint:* Final cost = Cost of bond + Accrued interest + Brokerage fee. Calculate time for accrued interest.)

Cost of bond: $\$1,000 \times .985 = \$985$

Time for accrued interest:

June	30 days
July	30 days
August	30 days
September	6 days
	96 days

Interest: $\$1,000 \times .0875 = \$87.50$

$\dfrac{\$87.50}{360} = \$.2430555$ per day
$\begin{array}{r} \times\quad\quad 96 \\ \hline \$\quad 23.33 \end{array}$

Final cost of bond	$ 985.00
Accrued interest	23.33
Brokerage fee	6.00
	$1,014.33

END-OF-CHAPTER PROBLEMS

DRILL PROBLEMS (*Note:* Problems for optional Learning Unit 22–3 follow the Challenge Problem 22–24, page 499.)

Calculate the mean (to the nearest hundredth):

22–1. $6, 9, 7, 4 = \dfrac{26}{4} = 6.50$

22–5. Calculate the grade-point average: A = 4, B = 3, C = 2, D = 1, F = 0 (to nearest tenth).

Courses	Credits	Grade	Units × Grade	
Computer Principles	3	B	9 (3 × 3)	
Business Law	3	C	6 (3 × 2)	
Logic	3	D	3 (3 × 1)	$\dfrac{43}{16} = 2.7$
Biology	4	A	16 (4 × 4)	
Marketing	3	B	9 (3 × 3)	
	16		43	

Find the mode:

22–9. 8, 9, 3, 4, 12, 8, 8, 9   8

22–13. Prepare a frequency distribution from the following weekly salaries of teachers at Moore Community College. Use the following intervals:

$200–$299.99
300– 399.99
400– 499.99
500– 599.99

$210	$505	$310	$380	$275
290	480	550	490	200
286	410	305	444	368

Salaries	Tally	Frequency
$200–$299.99	|||||	5
300– 399.99	||||	4
400– 499.99	||||	4
500– 599.99	||	2

22–15. How many degrees on a pie chart would each be given from the following?

Wear digital watch	42%	.42 × 360° = 151.2°
Wear traditional watch	51%	.51 × 360° = 183.6°
Wear no watch	7%	.07 × 360° = 25.2°
		360°

WORD PROBLEMS

22–17. The March 20, 2000, issue of *Industry Week/IW* reported on the latest interest yields. United States interest rates were compared to European Union rates. From the following information, prepare a line graph comparing the United States and European Union:

	Latest	**3 months ago**	**6 months ago**	**12 months ago**
USA	6.59%	6.75%	6.50%	6.20%
European Union	5.56	5.30	5.40	5.50

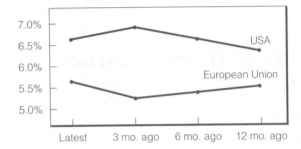

22–19. Bill Small, a travel agent, provided Alice Hall with the following information regarding the cost of her upcoming vacation:

Transportation	35%
Hotel	28%
Food and entertainment	20%
Miscellaneous	17%

Construct a circle graph for Alice.

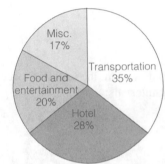

$.35 \times 360° = 126°$
$.28 \times 360° = 100.8°$
$.20 \times 360° = 72°$
$.17 \times 360° = 61.2°$

CHALLENGE PROBLEM

22–23. On June 15, 2001, the *San Francisco Business Times* reported on the annual revenue of the largest travel agencies in the Bay Area. The results are as follows

AAA Travel Agency	$86,700,000
Riser Group	63,200,000
Casto Travel	62,900,000
Balboa Travel	36,200,000
Hunter Travel Managers	36,000,000

(a) What would be the mean and the median? **(b)** What is the total revenue percent of each agency? **(c)** Prepare a circle graph depicting the percents.

a. $\dfrac{\$285,000,000}{5} = \$57,000,000$ mean

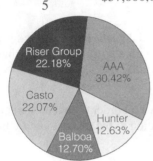

c.
$.3042 \times 360° = 109.51°$
$.2218 \times 360° = 79.85°$
$.2207 \times 360° = 79.45°$
$.1270 \times 360° = 45.72°$
$.1263 \times 360° = 45.47°$ $\$62,900,000 = $ median

b.

AAA $\dfrac{\$86,700,000}{\$285,000,000} = 30.42\%$

Riser Group $\dfrac{\$63,200,000}{\$285,000,000} = 22.18\%$

Casto $\dfrac{\$62,900,000}{\$285,000,000} = 22.07\%$

Balboa $\dfrac{\$36,200,000}{\$285,000,000} = 12.70\%$

Hunter $\dfrac{\$36,000,000}{\$285,000,000} = 12.63\%$

END-OF-CHAPTER PROBLEMS

DRILL PROBLEMS

1. Calculate the range for the following set of data: 117, 98, 133, 52, 114, 35.
    $133 - 35 = 98$

3. 7, 3, 12, 17, 5, 8, 9, 9, 13, 15, 6, 6, 4, 5
    Sum of the squared deviations = 237.5
    $237.5 \div (14 - 1) = 18.26923077$
    $\sqrt{18.2693077} = 4.3$ standard deviation

WORD PROBLEMS

5. The mean useful life of car batteries is 48 months. They have a standard deviation of 3. If the useful life of batteries is normally distributed, calculate **(a)** the percent of batteries with a useful life of less than 45 months and **(b)** the percent of batteries that will last longer than 54 months.
    a. $100\% - (50\% + 34\%) = 16\%$ **b.** $100\% - (50\% + 34\% + 13.5\%) = 2.5\%$

9. The time in seconds it takes for 20 individual sewing machines to stitch a border onto a particular garment is listed below. Calculate the mean stitching time and the standard deviation to the nearest hundredth.

67	69	64	71	73
58	71	64	62	67
62	57	67	60	65
60	63	72	56	64

    $1,292 \div 20 = 64.6$ mean time
    The sum of the squared deviations equals 478.8
    $478.8 \div (20 - 1) = 25.2$
    $\sqrt{25.2} = 5.02$ standard deviation